ABORTION
SCIENCE AND SPIRITUALITY

The Viewpoint of a Spiritist Physician

FABIO ARÁUJO

ABORTION
SCIENCE AND SPIRITUALITY

The Viewpoint of a Spiritist Physician

Spiritist Publishing GROUP

spiritist Federation of Florida

Original Title in Portuguese:
Aborto: Ciência e Espiritualidade

Translated by Darrel W. Kimble and Marcia M. Saiz

First Edition – 4/2026

ISBN 978-1-953672-85-8

SPIRITIST FEDERATION OF FLORIDA
Miami Beach, Florida (USA)
www.iamspiritist.us
info@spiritistfederation.us
(305) 496-6504

By purchasing this book, you are collaborating with supporting the Spiritist Movement in order to spread Jesus' Gospel under the guiding light of Spiritism.

INTERNATIONAL DATA FOR CATALOGING IN PUBLICATION (ICP)

A663a Araújo, Fabio. 1984–

Abortion. Science and Spirituality. – Translated by Darrel W. Kimble and Marcia M. Saiz. – First edition – Miami, FL (USA): 2026.

108 p.; 23cm

ISBN: 978-1-953672-85-8

1. Spiritism. 2. Abortion. 3. Social Conduct. I. Title

CDD 304.667

DEDICATION

A basketball coach was once questioned about allowing one of his players to attend the birth of his first child on the very day of a decisive game for his team. Wisely, the coach asked the journalist if he had children, and the answer was no. The coach then told him that when he had his first child, he would understand because he would be experiencing the pinnacle of human experience.

In fact, I discovered the purest and most genuine form of love with the birth of my firstborn. She is my greatest inspiration to try to be a better person every day.

So, it is with a love that cannot be contained in my heart, that I dedicate this work to you, my beloved daughter, Maria Eduarda.

CONTENTS

PREFACE

Life faces threats of destruction by some who, though well-endowed intellectually, demonstrate fragmentation in their overall evolution, in that they are ignorant of or prejudiced against the spiritual dimension of the human being, thus displaying an ethic based on atheistic materialism, and failing to include the manifestations of life on its transcendent axis.

Blinded to that which transcends the physical realm, they focus entirely on the energy of sensory life, which include, at most, macroscopic and microscopic life; they are incapable of glimpsing life in its full power, beyond what the biological body perceives, even with the help of devices.

Consequently, they deny the spirituality presented by religious individuals. They do not consider dark energy, which some astrophysicists suggest as being necessary in order to understand the universe. They reject any notion of transcendent life, widely represented nowadays by NDE (Near-Death Experience) research, Visions of the Dying, Past Life Therapies, Parapsychological Theta Phenomena, etc., in addition to the beautifully robust scientific-philosophical contributions of Spiritism, presented by Allan Kardec.

They are activists of material ecology, seeing only the front and not the back, i.e. that which is hidden and which controls biological existence; thus they lose sight of the perspective of the (uni)verse, of the multiple universes, in short, cosmic unity.

The zygote is the raw material shaped by morpho-bio-modeling energy, that is, by the thinking energy, which acts using the laws of genetics to structure embryonic and fetal life.

All students and health professionals, especially gynecologists and obstetricians, anesthesiologists and urologists, embryologists and geneticists, should study the post-materialist existential paradigm during their professional training. They should also undergo intrauterine memory regression to experience for themselves what the new epistemological model says about the life of the unborn child.

Those who regress to experience their own intrauterine life will never consider the neuroblast,[1] around 14 days of gestation, as the marker of the beginning of life, nor will they continue to assume that there is no intricately constituted pulsating mental life in the fetus inside the womb.

At any stage of pregnancy, the fetus is capable of perceiving the world outside the mother's body; it can reproduce dialogues between the parents or between the mother and other individuals; it can experience maternal emotions and thoughts that will affect it positively or negatively later after birth; it can describe in detail the mother's external environment; and, above all, it can be traumatized by dysfunctional maternal emotions... In short, it is a witness and co-participant in everything the mother experiences in her daily life. This has been my experience in my daily professional life during therapeutic regressions to the womb.

***

This book is a detailed and in-depth analysis of the entire reality to which I have referred, presented by a qualified and competent professional who does so with mastery and clarity.

Fábio Araújo Jr., a medical specialist in radiology dealing with images of life in the maternal sanctuary — the uterus — with training in various fields of psychology, including psychiatry and memory regression, is qualified, not only because of his academic training, but also because of his studies and practices in the field of spirituality, to discuss the subject with the authority of someone who has mastered the different aspects of the human being, thereby presenting a comprehensive view of the life of the child before birth.

A physician and therapist, enthusiastic about life in its many manifestations in nature, he has chosen to present us with this book, a true tribute and expression of respect for the infant coming into the world at a time of profound crisis and disagreement about the value of the unborn child's life.

---

[1] Neuroblast: An embryonic cell from which a nerve cell develops. http://www.thefreedictionary.com/neuroblast. – Tr.

His writing meets the academic public's thirst for knowledge, as well as that of the religious and secular spheres in general. With his easy-to-understand writing style, he has managed to reveal hidden truths in a simple manner, despite the seriousness of the subject matter.

No one will remain "on the fence" after reading this work, which I highly recommend.

ALBERTO RIBEIRO DE ALMEIDA
(Medical Doctor and specialist in Homeopathy and Systemic Family Therapy; postgraduate degrees and certificates in transpersonal psychology, psychiatry, family constellation, Neuro-Linguistic Programming (NLP), Past-life Therapy (PLT) and energetic dynamics of the psyche.

INTRODUCTION

There is a need to awaken humanity to a cry, a sob, a voice that is continually silenced, one whose reverberations are muffled in a world filled with easy answers for complex problems, where materialism and hedonism reign supreme, leading human beings to a disconnect from the ideals of morality and spirituality, to the point of trivializing existence itself, exterminating the possibility of life for thousands of tiny, defenseless babies every year.

From the Latin *Aboriri*, abortion means the premature termination of the fetus's development and its subsequent expulsion from the mother's uterus. In simple terms, we can divide the types of abortion into two main groups: spontaneous and induced.

Spontaneous abortion [miscarriage], which occurs naturally, is highly prevalent. Globally, about 15% of pregnancies end in spontaneous abortion, with most occurring during the first 12 weeks. On the other hand, in induced abortion, the fetus is expelled as a result of the intent of the mother or other individuals involved in the pregnancy. This type of abortion is performed using medication or mechanical means, such as vacuum curettage — removal of the conceptus by inserting a hollow curette into the uterus, through which a vacuum is applied.

A phenomenon present throughout history, especially due to the prevalence of miscarriages, abortion has been both accepted and rejected in various cultural contexts and has gained much importance as a topic of heated discussion in contemporary society, especially as of the latter half of the 20th century.

In Ancient Greece, Spartans sacrificed children with physical disabilities by throwing them off a cliff because they would not be productive members of society, and Aristotle believed that to avoid overpopulation, mothers could abort their offspring before they had "feeling and life."

In the sequence of Greek history, we have the clear stance of Hippocrates, the father of medicine, in favor of life when he states, "I will not give any woman an abortive substance," and in the passage that is part of the medical oath, "I WILL MAINTAIN the utmost respect for human life from the time of conception; even under threat, I will not use my medical knowledge contrary to the laws of humanity."[2]

From a religious perspective, abortion is generally contraindicated, based on the understanding that there are divine purposes; that life begins at conception, and the dual view of our existence (we are matter and spirit). Religions usually make exceptions in cases where there is a need to save the mother's life, which is therefore a morally acceptable stance from a religious point of view.

In 1798, Thomas Malthus, in his book *An Essay on the Principle of Population*, indicated that the population could grow to a point where there would not be enough food for everyone, thus giving rise to a state of social collapse. Malthus advocated sexual abstinence to curb population growth. Since he was a pastor, he did not advocate abortion, but he did end up directly influencing the pro-abortion movement in the following centuries with that particular argument, which did not materialize after all, since technological evolution substantially increased the world's food production capacity. The biggest problem today is the unequal distribution of food, which, in actuality, is produced in sufficient quantities to easily fill the refrigerators of the entire human population.

The 20th century, considered by many to be "The Age of Narratives," was marked by important sociopolitical movements that directly impacted the discussion surrounding abortion. One of the main ones was the Neo-Malthusian movement, with a geopolitical vision concerned with population growth, especially in underdeveloped countries, believing that this would bring poverty and unemployment, in addition to consuming the resources of "developed humanity." In 1968, in his book *The Population Bomb*, author Paul Ralph Ehrlich estimated that starting in the 1970s, we would be unable to feed the world's population, creating an apocalyptic scenario and influencing much of the developed world with fearmongering. This moment was so delicate that U.S. President Lyndon Johnson stated in 1965 that it would be better to invest $5 in population control than $100 in the development of poor countries.

---

[2] English text of the *Declaration of Geneva*, 1948. Available online at: https://www.wma.net/wp-content/uploads/2016/11/Decl-of-Geneva-v1948.pdf. – Tr.

With this in mind, powerful foundations, such as the Rockefeller Foundation, dedicated themselves to studying and developing plans for population control, while the United Nations (U.N.) developed policies to promote birth control and resource control.

In 1974, the U.S. government published *The Kissinger Report*,[3] which warned of the depletion of natural resources and conflicts of interest between developing countries due to population growth and increased industrial production. It recommended that population control assistance be employed in larger and faster-growing developing countries where the U.S. had special political and strategic interests, such as India, Pakistan, Nigeria, Turkey, Colombia and Brazil. The report emphasized the importance of not being seen as a policy by industrialized countries aimed at using resources and reducing the power of developing countries.

The *Kissinger Report* was categorical in its defense of abortion in the context of population and resource control, stating that "no country has reduced its population without resorting to abortion."[4] Consequently, in order to mask the true objectives (reducing the birth rate and preserving resources), the Family Planning concept began to be used.

During this same period, the feminist movement, which fought for women's civil and political rights (the right to vote, education, and entry into the labor market), embraced the idea of "family planning" and began to fight for the legalization of abortion.

One of its forerunners, Margaret Sanger, had gained recognition by promoting and defending ideas about abortion, divorce, contraception and eugenics. Sanger, a member of the American Eugenics Society, stated that: "Eugenics…is the most adequate and thorough avenue to the solution of racial, political, and social problems."[5] "… the most urgent problem today is how to limit and discourage the over-fertility of the mentally and physically defective.."[6]

---

[3] National Security Study Memorandum, NSSM 200, Implications of Worldwide Population Growth for U.S. Security and Overseas Interests, *(The Kissinger Report)* Dec. 10, 1974. Available at: https://archive.org/details/pcaab-500-3/PCAAB500-3/mode/2up?q. – Tr.

[4] *Kissinger Report*, 114. – Tr.

[5] *Congressional Record*, Vol. 155, No. 54, p. 1. Available at: https://chrissmith.house.gov/uploadedfiles/2009-03-31_in_awe_of_a_eugenicist.pdf. – Tr.

[6] Margaret Sanger. "*The Eugenic Value of Birth Control Propaganda.*" Birth Control Review, October 1921, page 5. Quote taken from the referenced website in the preceding footnote, p. 4. – Tr.

In addition to considering pregnancy a disease for women, Sanger believed that "The most merciful thing that a family does to one of its infant members is to kill it."[7] Sanger opened her first abortion clinic in a black neighborhood in New York City, and years later she would participate in the world's largest abortion foundation, *Planned Parenthood*, which in recent decades has become the largest abortion institution in the world, performing around 400,000 abortions in the U.S. in 2020, leading the organization to the impressive mark of $2 billion in net assets, according to data reported in its own balance sheet.

Planned Parenthood is accused of being responsible for African American genocide in the U.S. Martin Luther King's daughter, Alveda King, reports that since 1973, when abortion was legalized, 10% of whites have been exterminated, whereas the number for blacks is 28%. The revealing American documentary *Blood Money* exposes, in part, the billion-dollar abortion industry and its consequences for the African American population, since currently in the U.S., out of every five black women who become pregnant, three choose to have an abortion, and about 1,500 African American children are aborted every day. It is no coincidence that 80% of abortion clinics in the U.S. are located in poor, black neighborhoods.

A contemporary of Margaret Sanger, English writer Marie Stopes, was part of the same movement defending abortion in Europe. Stopes also advocated the compulsory sterilization of "inferior races" and eugenics. She was responsible for founding the first family planning clinic in the United Kingdom, as well as founding Marie Stopes International, an abortion organization that operates in more than 37 countries, mostly in Africa, with the support of the U.N.

Represented by a highly lucrative billion-dollar industry and a contemporary cultural movement linked to the sexual revolution of the 1960s, it is known that, since the end of the last century, abortion has been funded by large corporations such as the Ford Foundation and the foundations of billionaires Bill Gates and George Soros, and being ever-present in political and media agendas, constantly modifying and reaffirming popular acceptance of the issue.

In this context, what was once unacceptable became debate worthy and gradually more acceptable. Today the idea of pro-abortion groups is to see the right to abortion as being a fundamental right of women.

In fact, the acceptance of abortion by a significant portion of society has enabled the United States, in the 50 years since its legalization in 1973, to reach

[7] See footnote no. 4 above. – Tr.

the terrifying estimate of 64 million abortions performed – millions of lives cut short while still in the womb, without the slightest chance of defense.

Another point that few refuse to discuss is that users of abortion clinics are usually from disadvantaged backgrounds and classes, and abortion is encouraged by many entities as an economic measure aimed at reducing the number of people dependent on government welfare services (Knowles, 2018).

Fundamentally, in addition to the aforementioned issues involving the abortion industry, women are extremely harmed by this procedure, given the consequences for their physical and mental health, as detailed in various scientific works and studies over the last few decades.

In an attempt to devalue the life of the developing baby, pro-abortion groups try to go against the fundamentals of biology and genetics, referring to the fetus as something non-human or without individuality; however, not only the basic concepts of science, but also serious recent studies, such as theories about fetal psychism, are crystal clear about the individuality of the pulsating human life in the womb.

Considering the fact that in Brazil there is a growing movement toward trivializing life and legitimizing abortion, it is necessary to make every effort to ensure that as many people as possible are aware of the risks and consequences of such a heinous act.

Our desire with this work is to open space for deep reflection on life, raising essential information about women's health in relation to abortion, defending the individuality of the developing being, highlighting how the union of science and spirituality can be the key to unveiling a new reality for humanity, and clarifying the spiritual context surrounding abortion through the lens of Spiritism.

CHAPTER 1
THE BEGINNING OF LIFE

"We must remember that wisdom is not just knowledge.
Wisdom is the ethical use of knowledge."
HERMÍNIO MIRANDA

"Human life begins at conception" is the categorical statement made by Karl Ernst von Baer, the father of modern embryology, defining the significance of the zygote, when a spermatozoon unites with an egg.

It is correct to say that the zygote contains all the genetic information necessary for the life of the new human being and its individuality, for it contains the complete set of chromosomes that carry the genes responsible for the physical characteristics and emotional tendencies of the future individual.

Although the basic laws of physics are capable of explaining many physical phenomena in the universe, they cannot explain the manifestation of human life.

Life is an emergent property of complex biological systems and cannot be reduced to simple physical laws; for example, it challenges the basic concepts of thermodynamics, where its second law (law of entropy) states that in a closed system disorder or randomness always increases over time. This means that energy tends to dissipate and become less organized over time. Life, on the other hand, is characterized by a reduction in entropy, that is, *by the creation of order from chaos*: from a single cell, a multicellular and extremely complex system is organized.

Fetal development is an incredible and fascinating process that begins at conception and continues throughout pregnancy. During this period, the embryo undergoes a series of amazing transformations. One of the first stages of fetal development is the formation of neural tissue, which occurs at the end of the first month of pregnancy. At six weeks, the embryo is sensitive to light; at seven weeks, it has well-defined brain and spinal cord structures, and by the 14th week, the fetus has well-defined hearing. In fact, fetal hearing and the importance of parental conversation during pregnancy are topics of interest in various scientific fields, including medicine, psychology and linguistics.

There are several important scientific studies related to this topic, such as:

- *"Prenatal Language Learning: What Do Babies Hear?"* (Moon et al., 2013): This study uses neuroimaging techniques to investigate how the fetal brain processes linguistic information. The authors show that fetuses are able to distinguish different speech sounds and that this ability may influence later language development.
- *"Maternal-Fetal Attachment and Fetal Heart Rate Patterns during Antepartum"* (Alhusen et al., 2013): This study investigates the relationship between maternal-fetal attachment and the fetal heart's response to auditory stimuli. The authors show that mothers who report greater attachment to their fetus have babies with a greater ability to adapt to sound stimuli.

Another important element in fetal development is the heartbeat. The heart begins to beat around the fifth week of pregnancy, and from that moment on, the heart is responsible for supplying oxygenated blood and vital nutrients to the developing fetus. The sound of the fetus's heartbeat can be heard in a simple ultrasound exam and can often be a determining factor in the first maternal-fetal psychological bond, and it frequently represents a turning point for mothers who are considering abortion.

The case of obstetrician-gynecologist Bernard Nathanson is a classic. Nathanson founded the National Abortion Rights League in the U.S., campaigning intensely for the legalization of abortion in that country in the 1970s. As director of the country's largest abortion clinic (the Center for Reproductive and Sexual Health), he was directly responsible for more than 60,000 abortions and personally performed more than 5,000, including one of his own offspring. According to his accounts, in the late 1970s he studied ultrasound to gain visual access to the fetus *in utero* and, as a result, suffered a personal crisis. Consequently, he radically changed his stance on abortion and became a pro-life activist. He produced the famous documentary "*The Scream of Silence*" (1984), in which he succinctly explains the details of the abortion procedure, highlighting that, at the moment of bodily perforation via the abortion equipment during the procedure, the baby opens its mouth, which Nathanson interprets as a scream, in addition to a vain attempt to escape the procedure inside the uterine cavity (Derosa, 2018; Bernard, 2019). Thus, Dr. Nathanson postulated that every woman should be required to have an ultrasound to see the fetus and hear its heartbeat before opting for an abortion.

Over the following weeks and months, the fetus continues to grow and develop, acquiring more physical characteristics and abilities. At around 16 weeks, the parents may begin to sense the first movements of the fetus, a vibrant human being with unique physical and personality traits.

In addition to the neuroanatomical issue with its sensory perceptions, we find in the psyche of this fetus the presence of perceptions of the environment, which go far beyond biological issues; this fact is very well documented in memory regressions that reach the intrauterine period and indicate important events that occurred at that time and which were fixed in the subconscious.

Research on fetal psychism confirms this hypothesis. One such study was performed by Dr. Alessandra Piontele and Dr. Antony de Casper, in which stories were told to fetuses inside the womb, and then, some time after birth, the stories were retold, using as a reference babies who had not heard the story when they were in the womb. Consequently, the electrodes recorded different emotional responses in the two groups, indicating that fetal psychism is indeed vibrant from very early on.

A study by DiPietro et al. (2006) investigated the development of fetal perception of external stimuli, such as sounds and vibrations, throughout gestation. The results showed that fetuses responded to sound and tactile stimuli, suggesting that they are capable of perceiving and processing information from the environment before birth.

Another interesting study was conducted by Kisilevsky et al. (2009), who investigated the effect of prenatal exposure to music on the emotional sensitivity of newborns. The results indicated that babies exposed to music during pregnancy had greater emotional sensitivity than those who had not been exposed.

There is also evidence suggesting that fetuses may exhibit distinct behavioral and emotional reactions in response to stimuli in the intrauterine environment. Some studies suggest, for example, that fetuses may have preferences for specific sounds and music, in addition to responding to visual and tactile stimuli, thus indicating individuality, considered by some authors to be an incipient personality.

Psychiatrist Alessandra Piontelli (1995) used serial obstetric ultrasonography to verify that from the 16th week of pregnancy until four years of age, fetuses have their own personality and that there is continuity in behavior before and after birth, despite the marked changes caused by birth and nature. Even more impressive are the reports of twin pregnancies in which it was evident that each fetus exhibited unique and striking behavior in relation to its twin.

In this context, understanding the existence of a fetal psychism, the demonstration of individuality in a genetically unique being, dismantles one of the main arguments of abortion advocates, that the fetus or embryo is nothing more than a cluster of cells, in an attempt to objectify a life that is fully active.

This modus operandi can be seen in various actions and references in the abortion industry, for example, when referring to the stage of the abortion procedure that involves crushing and removing the fetal skull from the mother's uterus as "the exit of number one," objectifying the most noble and well-protected structure of the body, the brain. Another strategy for avoiding contact with reality, namely that there is a fully developing human being in the womb, is the reference by the so-called progressive political-cultural movement to the embryo/fetus as a cluster of cells, thereby diminishing life in a certain way and making the ideals of abortion more palatable.

It is worth remembering that history is permeated, from time to time, by discourses denying humanity and/or affirming sub-humanity with regard to certain people, as was done over the centuries with indigenous and black people; or how Adolph Hitler managed to legitimize the persecution of Jews as an inferior race, transforming their lives and individualities into mere disposable numbers.

Furthermore, it is important to emphasize the fact that, once legally permitted, abortion can be an open door for the application of an invisible and undeclared selection called "eugenics."

Eugenics is a term that refers to the search for the improvement of the human race through the selection of characteristics considered desirable and the elimination of those considered undesirable.

The word eugenics (from the Greek eu-, "good," and -genéia, "evolution," "origin," "race") means "good lineage." The "science of eugenics," known as the "science of good breeding," was developed by Francis Galton (1822-1911) in England, influenced by his reading of the book *The Origin of Species* (1859) by his cousin, Charles Darwin. Historically, the term has been associated with racist ideologies that seek to promote the superiority of certain races or ethnic groups over others.

In practice, eugenics can take many forms, including forced sterilization of those considered "undesirable" or "inferior," incentives for people considered "superior" to have more children, or the selection of partners based on race or physical characteristics.

One of the main examples is Down syndrome, historically considered an "undesirable" characteristic and used as justification for discriminatory policies and practices, such as in Iceland, where abortion is legal, and practically no children with Down syndrome are born anymore – they are all murdered before birth. Another example is the United Kingdom, where the number of abortions related to Down syndrome is close to 90%.

On the other hand, many parents who have children with Down syndrome report that the experience has profoundly changed their lives for the better. Although there may be challenges and difficulties, many parents describe how their children with Down syndrome bring joy, love and a unique perspective on life. They learn to appreciate the simple things and live in the moment, and often develop a greater appreciation for the little things in life.

Around the world, organizations led by individuals with the syndrome and their families strongly oppose the selective abortion of fetuses with Down syndrome. They argue that selective abortion based on the presence of a disability is discriminatory and sends a harmful message about the value of human life. These organizations hold that all human lives have intrinsic value and deserve respect and protection, regardless of disability or medical condition. They also argue that Down syndrome should not be viewed as a sentence of limited life or poor quality of life, as many people with Down syndrome live happy and productive lives.

Therefore, it is necessary to embrace the idea that, from conception, and from every perspective, human life is a non-negotiable asset of inestimable value, which must be protected and safeguarded by all of us.

CHAPTER II
HEALTH AND PROTECTION OF THE WOMAN

"Let not your heart be troubled. Believe in God, believe also in me."
JESUS IN JOHN 14:1

In the current context of narratives without practical or scientific foundation, pro-abortion movements argue that such a procedure would be a woman's civil right and a matter of public health, stating that the legalization of abortion would be, above all, a means of reducing maternal mortality and the psychological consequences of an unwanted pregnancy.

As it turns out, contrary to what pro-abortion activists claim, legalizing abortion is not an initiative to improve public health for numerous reasons.

First of all, it is not intervening in a disease; furthermore, and even more importantly, in order to justify support for its legalization, the relationship between abortion and maternal mortality rates is often concealed. The figures are alarming with several studies showing that the risk of maternal death is 10 times higher up to the 10th year after an induced abortion. The editors at Williams Obstetrics point out that maternal death from induced abortion is 14 times higher than in cases where the pregnancy continued, which shows that pregnancy is protective of maternal health (Cunningham et al., 2016).

Moreover, legalizing abortion increases the occurrence of complications, especially maternal mortality. The experience of countries that have passed laws permitting abortion shows that the problem of abortion complications has not been resolved.

When it comes to pregnancy and public health in Brazil, the greatest concern should be with the main causes of complications that are directly linked to maternal mortality: hemorrhage, hypertension and infection. These three pathologies have historically high rates in our country, which is due to the difficulty the pregnant population has in receiving adequate care throughout the prenatal period and actual assistance during childbirth.

Indeed, this should be the focus of the general population's demands, valuing women and life above all else. If the public health system cannot even provide minimally satisfactory care to the population during pregnancy, imagine adding to the public coffers the high costs of medical care for abortion procedures and their complications.

The situation becomes even more delicate considering the obvious increase in the number of abortions in the years following legalization, reaching a plateau well above the initial rates. As a practical example, we have the increase in abortion rates of 1,200% in Mexico, 512% in Spain, and 169% in Portugal.

In an attempt to disguise reality, the abortion industry uses fictitious figures, creating numbers and statistics without scientific basis. In the U.S., before the legalization of abortion, the pro-life group indicated that there would be around 200,000 procedures annually, while the pro-choice entities estimated 1,000,000 procedures; after legalization, in the first year, around 170,000 abortions occurred (a number very close to the pro-life group's estimate).

On this subject, Dr. Nathanson states:

> I confess that I knew the figures were totally false, and I suppose the others did too if they stopped to think about it. But in the "morality" of our revolution, it was a useful figure, widely accepted, so why go out of our way to correct it with honest statistics? The overriding concern was to get the laws eliminated, and anything within reason that had to be done was permissible.[8]

This modus operandi has occurred in several countries, such as Uruguay, where pro-abortion activists reported 33,000 abortions per year, a figure disputed by pro-life advocates (indicating a blatant overestimate); in the first year after

---

[8] Bernard Nathanson, *Aborting America.* (Garden City: Doubleday & Co. 1979), 193. – Tr.

legalization, the actual number was 4,000 abortions. After legalization, the numbers rose annually and stabilized at around 8,000 per year.

In Brazil, it is no different. The figures reported in the media have no basis in fact or evidence. According to calculations reported in two studies conducted without technical and statistical criteria, the estimated number of abortions in the country would be 1,500,000, with a probable rate of 11,000 deaths per year from complications, all based on a study by the magazine *Abrasco*, an actively pro-abortion entity linking an absurd number of 1 death for every 5 illegal abortions performed in the country (without presenting the data to support this conclusion).

The reality is completely different from the estimates in the above studies. The number of maternal deaths from illegal abortions is much lower, less than 100 cases per year, according to data from the Ministry of Health. And contrary to what many people think, if abortion were legalized, the trend would be for maternal deaths to increase in proportion to the increase in the number of abortions (even if they were performed in hospitals).

The physical and psychological repercussions of induced abortion are often devastating – the list is long, and the pain is great. Some consequences manifest immediately, but many only appear years later. The female reproductive system is clearly shaken by the procedure and may suffer more immediate repercussions from infections and hemorrhaging, which may lead to the need for surgical removal of the uterus and, in more severe cases, death. The chances of endometritis (chronic inflammation of the endometrium), ectopic pregnancy (a delicate condition of pregnancy outside the uterine cavity, requiring surgical intervention), and infertility as a later complication increase significantly.

The relationship between breast cancer and induced abortion was established many years ago. Daling et al. (1994) found a 30% increase in American women, and when induced abortion occurred in the first pregnancy, the increase was 50% because pregnancy protects against breast cancer (Macmahon et al., 1970). In the case of abortion in the first pregnancy, the mammary lobe undergoes incomplete maturation, making it more vulnerable. The literature review conducted by Brind (2015) showed that this association is robust, as did the Chinese study by Huang et al. (2014), which evaluated a representative sample of the Chinese population and found a strong relationship between induced abortion and breast cancer.

Unfortunately, complications are not limited to the reproductive system and breasts. A correlation has been established between the progression or onset of autoimmune diseases, such as rheumatoid arthritis, multiple sclerosis and thyroiditis, contributing to conditions that are often protracted and difficult to treat.

In order to receive the child, there is a whole process of organic preparation, which corresponds to a parallel psychological process. After the abortion, hormones continue to circulate in the body for some time, contributing to physical and psychological emptiness, unlike in a full-term pregnancy, where mother and child are ready to separate.

The physical body becomes the mouthpiece for the protests of a woman's psyche, often generating physical repercussions (somatizations), according to a study conducted in 2018 by Petersen and collaborators.

Therefore, given the irrefutable data related to the physical complications of induced abortion, especially the fact that maternal mortality increases significantly compared to women who carry their pregnancies to term, advocates for women should, more than anyone else, protect them by discouraging abortion.

CHAPTER III
PSYCHO-EMOTIONAL CONSEQUENCES OF ABORTION

"Unexpressed emotions never die. They are buried alive and come out in the worst forms later."
SIGMUND FREUD.

In addition to the physical symptoms resulting from abortion, there are also significant psychological and emotional consequences for the woman's health.

Delicate psycho-emotional repercussions have been studied in recent decades in patients who have undergone abortion procedures, with anxiety disorders, depression, substance abuse, and their immediate and delayed complications as the basic manifestations, as pointed out in a scientific study by Fergusson, Horwood, and Boden (2008). They studied 500 women up to 30 years of age and concluded that those who had induced abortions had 30% more mental disorders than those who had not undergone the same experience.

Some authors argue that women do not seek abortion as a method of "contraception," but rather to terminate unwanted pregnancies resulting from unwanted sexual relations, even when these were not the result of coercion or sexual violence.

In this sense, arguing that an unwanted pregnancy can cause more psychological damage than induced abortion is to ignore scientific knowledge and attempt to cover up the epidemic of depressed women, who can ill afford to admit the cause of their illness, thereby eclipsing their guilt and preventing the resolution of the problem simply because society accepts abortion.

According to a scientific article published in a British psychiatry journal, there is a 34% increase in anxiety levels in women who have undergone induced abortion, a 110% increase in alcoholism, and a 155% increase in the number of suicides.

One of the most important studies on induced abortion is that of Broen et al. (2005), which was interdisciplinary in nature and conducted by the departments of Behavioral Sciences, Gynecology, and Obstetrics at Buskerud Hospital and the Institute of Basic Medical Sciences at the University of Oslo (Norway). It is worth noting that during this research, Norway was the world leader in the Human Development Index (HDI - score 0.928). In that country, abortion has been permitted up to the 12th week of pregnancy since 1978 (the government even sponsors abortion services), and there are approximately 9,000 miscarriages and 15,000 induced abortions annually, demonstrating how abortion is considered legal and morally acceptable.

The researchers assessed the impact of abortion on the mental health of 120 Norwegian women ten days, six months, two years, and five years after the abortion, investigating their quality of life and the occurrence of anxiety and depression. It is noteworthy that the follow-up of these women was quite long (five years), enough time to assess the abortion-related emotions and their short-, medium-, and long-term effects. The results revealed that women who suffered miscarriages had higher depression and anxiety scores in the short term (at 10 days and 6 months), whereas the induced abortion group had significantly higher scores than the first group and the general population in the long term (at 2 and 5 years), with a predominance of feelings such as denial, guilt, and loss. The induced abortion group stood out for its high levels of anxiety, depression and poorer quality of life than that of the general population.

Scholars point to the existence of a guilt complex, which, even if unconscious, would be decisive for some degree of mental illness among those involved, manifesting itself in various ways, especially in women, who may experience nightmares, depression, fear of dying, and fear of no longer being able to have children. The most serious cases occur when they are unable to forgive themselves and begin to adopt an attitude of self-condemnation towards life, unconsciously seeking situations of self-sabotage that allow them to respond to the calls of toxic guilt.

Feelings of loneliness and loss are also common when the woman, not feeling comfortable with sharing, ends up experiencing solitary mourning due to apparent helplessness.

Furthermore, even in cases where no negative emotional repercussions arise at first, in the long term, the shame, regret and guilt due to the reprehensible act follow one another like overwhelming inner storms.

Post-traumatic stress disorder is a chronic condition that can occur after emotional trauma and can last for months or even years. It can be triggered by events that bring back memories of the trauma and cause intense emotional and physical reactions. Symptoms of post-traumatic stress disorder include nightmares, flashbacks, avoidance of situations that remind the person of the trauma, exaggerated reactions to stimuli, anxiety and depressed mood.

In the case of abortion, some individuals may experience similar symptoms known as post-abortion syndrome. This syndrome is characterized by nightmares, bitter memories of the experience, aversion to baby-related items, and other emotional and physical reactions that can persist for a long time.

Mota, Burnett and Sareen (2010) studied a group of women representative of the Canadian population; suppressing other causes, legal induced abortion in the country increased the risk of mental disorders: mood disorders (61%), social phobia (61%), suicidal ideation (59%), alcohol abuse (261%), alcohol dependence (142%), drug dependence (287%) and drug mixing (313%).

In Russia, where abortion has been legal for over 100 years, inducing abortion has led to serious ego disruption, according to a study by Keenan et al. (2014), and is linked to alcoholism, which in turn has led to further unwanted pregnancies and abortions. In this succession of events harmful to the psyche, alcoholism became more severe, and abortions became recurrent. This is a real calamity, as about 60% of Russian women have had an abortion.

Even more critical are suicides among the Finnish population (Gissler; Hemminki; Lonngvist, 1996). These authors studied population-based data on suicide deaths according to reproductive outcomes (childbirth, miscarriage and induced abortion) between 1987 and 1994, comparing them with the general population. They found that the group with the highest number of suicides was women who induced abortions.

Indeed, abortion violates a law of life and nature, and perhaps for this reason, women are more vulnerable to mental disorders and suicide. It is a matter of humanity; pro-abortion movements must therefore understand that defending the life of the baby is also defending the life and mental health of women.

CHAPTER IV

ABORTION – THE DARKNESS

"Whoever commits an injustice is always more unhappy than the one who is wronged."
PLATO.

The abortion procedure may vary depending on gestational age. In general, in the first two months (up to 8 weeks), the procedure is performed via medication induction, especially mifepristone, which induces the death of the embryo, and two days later misoprostol, which promotes uterine contractions until the embryo is completely expelled. If this is not successful, manual vacuum aspiration is performed using a long, thin tube inserted into the uterus, and the embryo is sucked out with a large, attached syringe.

From the second trimester of pregnancy onwards, the fetus is too large to pass through the suction tube, so the cervix has to be dilated, which is done with numerous thin rods (one or two days before the procedure). Then the doctor removes the fetal parts with forceps; the skull of the fetus is crushed to facilitate removal and a sharp tool (called a curette) is used to scrape out the contents of the uterus to remove any remaining tissue.

The most dramatic situation occurs after 20 weeks, when the abortion procedure lasts an average of three days. During the first two days, the cervix is dilated using rods and medication is administered for pain. On the third day, the abortionist uses ultrasound to locate the fetus's legs. Holding on to one leg with forceps, the doctor pulls the fetus out up to the head. Next, scissors are inserted into the base of the skull to create an opening. A suction catheter is placed in the opening to remove the brain. The skull collapses and the fetus is removed.

KCI is often injected into the fetus's heart to establish fetal death before the procedure described above. This injection can be performed without anesthetics, something which is of unprecedented cruelty, since even when performed on an animal for euthanasia, anesthesia is administered before the injection. It is also known that the embryo at around five weeks already has peripheral nerve endings similar to those of an adult. Dr. Colen Malowey, a neonatologist at Northwestern University, states that in her intrauterine procedures, such as venous access, chest drainage, lesion biopsy, etc., fetuses feel and react to the manipulation; just imagine in an abortion procedure, when there is separation of limbs and crushing of the skull.

The medical course, which must be based on ethical and moral principles of beneficence, respect and humaneness, traditionally ends with the Hippocratic Oath (Hippocrates is considered the father of medicine). An excerpt from this oath reads: *"I WILL MAINTAIN the utmost respect for human life from the time of conception; even under threat, I will not use my medical knowledge contrary to the laws of humanity."9*

Thus, when health professionals sell out to the "death industry," and disconnect from what is most genuine in their professional practice, they end up unconsciously placing themselves in vulnerable situations regarding the psychological aspect, experiencing states of anxiety, depression and guilt in the medium term, according to a study by Buluk and O`Neil (1980).

There is no way to understand as normal or healthy a society that advocates the extermination of a baby in the womb, especially at a gestational age when it could already live outside the uterus. From 20 weeks onwards, and especially when the fetus already weighs around 500 grams [17.6 oz.], the chance of surviving a premature birth is enormous. Thus, in such cases, if the expulsion of the child from the womb is induced, it will be born alive. There is an organization in Brazil that cares for children who were born alive after abortion attempts, many of them with sequelae from the unfortunate attempt.

There is no medical justification for feticide when considering the procedure for terminating a pregnancy. In 2020, a baby weighing just 212 grams [7.5 oz.] was born prematurely in Singapore (at just 22 weeks of gestation). After 13 months in the hospital, the healthy baby was discharged weighing just over six kilograms [13 lbs.].

---

[9] *Declaration of Geneva*, op. cit. – Tr.

When it comes to justifications, we find in the feminist movement claims that abortion is a lesser evil, as if we could relativize life. What values would be greater than human life itself? There is talk of a woman's right over her own body, thus forgetting that in pregnancy there are two bodies, and there can even be two female bodies. This argument maintains the individualistic and selfish use of one's own body; it is further abusive in the sense that it relies on the denial of the body of the unborn child, who is absolutely defenseless and incapable of protesting and resisting the violation of its right to live.

Despite the notable contributions that this movement has made to securing women's citizenship rights, the proposals put forward are based on a materialistic understanding, opening the door to ideas such as those of Margaret Sanger (mentioned in the Introduction), the 20th-century feminist with racist and eugenicist views, who claimed that the way to create a pure-blooded race and rid the world of human parasites would be to legalize and encourage abortion. It is important to remember that currently 3 out of 4 abortions in the U.S. are performed in low-income neighborhoods on African American women, which is one of the main reasons for the decline in that population's birth rate in the country.

No less repugnant is the abortion performed based on the baby's sex, as in South Korea, India and China. When the baby's sex is discovered to be female, the abortion is performed at the woman's discretion, regardless of justification. It is estimated that around 7 million female babies are aborted each year in India – nearly 14,000 per day.

The issue of abortion reaches dramatic proportions when it involves the worst form of aggression: rape. This heinous act of violence can leave profound and terrible emotional scars on the victim and can be a precursor to severe anxiety disorders. The victim deserves full support and compassion, given the gravity of the crime. The disgust inherent in the act itself is often transferred to the possible embryo conceived through violence, and which is also seen as an aggressor, but which, in fact, is another victim, a silent one, unrecognized by the justice system and incapable of exercising the basic right of a human being – the right to live.

Sent "to death row" with the authorization of the law in most countries (including Brazil[10]), the baby is forgotten in the name of the violence in which

[10] Currently, in Brazil abortion is legal in three situations: 1) rape; 2) risk to the woman's life; 3) fetal anencephaly. https://www.gov.br/saude/pt-br/assuntos/saude-de-a-a-z/s/saude-da-mulher/saude-sexual-e-reprodutiva/interrupcao-gestacional-prevista-em-lei. – Tr.

it was conceived, and is condemned as a criminal without having done anything against any of those involved in the drama.

The psychological damage resulting from the assault does not disappear with the abortion; on the contrary, new traumas may arise, as discussed in the previous chapter.

The position of the Spiritist Medical Association[11] is viscerally opposed to abortion, even though it is supported by human legislation. Comprehensive treatment should be provided to the mother during and after pregnancy, while at the same time safeguarding the life of the child inside and, later, outside the womb.

Therefore, the ideal and healthiest option for the mother (knowing the physical and psychological repercussions, including increased maternal mortality) and, obviously, for the baby, is to carry the pregnancy to term and give the child up for adoption, if the mother does not want to be or cannot be a mother after the birth.

In Brazil, we have what is known as "voluntary surrender," which consists of the possibility of a pregnant woman or mother giving up her child or newborn for adoption in a procedure assisted by the Children and Youth Court, guaranteeing and including absolute confidentiality so as not to expose those involved.

Dr. Jérôme Lejeune (member of the French Academy of Medicine) thoughtfully makes three interesting points:

- If most legislation in the West does not condemn rapists to the death penalty, why condemn an innocent being to such an end?
- If a zygote is not a human being, how would it become one if nothing is added to it?
- A fetus is a patient; medicine is designed to heal. Any technical, moral or legal discussion is superfluous; we simply have to choose between medicine that heals and "medicine" that kills.

[11] AME – Associação Médica Espírita do Brasil. – Tr.

CHAPTER V

A NEW PARADIGM IS PRESENTED

"We are not humans having a spiritual experience; we are spiritual beings having a human experience."
TIELHARD DE CHARDIN

With scientific and technological advances, human beings have been able to walk on the moon and detect what happens millions of light years away from our planet; in addition, artificial intelligence is opening up an unimaginable range of solutions and possibilities. In the microcosm, knowledge of subatomic particles has unveiled a new reality, altering physics' centuries-old structures of reasoning. For example, based on its experiments, quantum mechanics establishes the observer (consciousness) as a determining factor in the behavior of matter.

On the other hand, despite all our intellectual progress, we have not yet been able to find our way within ourselves. The thinker Thomas Hardy tells us that humanity has lost God's address, while Divaldo Franco asserts that nowadays, humanity has lost its own inner address.

Emotional problems are multiplying due to a loss of personal identity, as well as a loss of connection with other people and life. Consequently, humanity has become a victim of individualism, sexism and consumerism, as pointed out by the eminent theologian and psychologist Rolo May.[12]

Amit Goswami, a world-renowned physicist whose studies seek to reconcile "science and consciousness," states:

> A critical level of confusion permeates the world today. Our faith in the spiritual components of life—in the vital reality of consciousness, of values, and of God—is eroding under the relentless attack of scientific materialism. On the one hand, we welcome the benefits derived from a science that assumes the materialist worldview. On the other hand, this prevailing worldview fails to satisfy our intuitions about the meaningfulness of life.[13]

According to the World Health Organization (WHO), mental disorders – especially depression and anxiety disorders – will be the leading cause of illness worldwide in the near future. Data show that currently one in four people has or will have mental disorders in the coming years; it is currently estimated that about 250 million people meet the criteria for a diagnosis of chronic depression.

Even more serious are the suicide rates, reaching the frightening annual estimate of 1,000,000 lives self-terminated.

In the midst of these delicate times, we see the mechanisms of escape from reality taking over society. In order to avoid facing the tormenting situation itself, modern human beings seek externally what they fail to construct internally; consequently, the feeling of existential emptiness is increasing.

---

[12] Rolo May (1909-1994) was an influential American existential psychologist and author, born in Michigan, who focused on the profound dimensions of human existence. After graduating from Oberlin College in 1930, he spent several years teaching in Greece and later became a counselor at Michigan State University. His academic journey led him to study theology and eventually earn a doctorate in counseling psychology from Columbia University in 1944. May's early works, including "The Meaning of Anxiety" and "Man's Search for Himself," explored the themes of anxiety, alienation, and the quest for self-understanding in a fragmented cultural landscape. https://www.ebsco.com/research-starters/history/rollo-may. – Tr.

[13] Amit Goswami, *The Self-Aware Universe* (New York: Jeremy P. Tarcher, 1993), 16. Found online at: https://www.thejungletimes.com/page/downloads/files/Amit%20Goswami-The%20Self-Aware%20Universe-Tarcher%20(1995).pdf. – Tr.

Dissatisfaction is responsible for much human suffering, generating imbalances that could easily be avoided. People use escape mechanisms to avoid accepting their own reality, creating fictitious models of happiness, onto which they transfer their aspirations, producing states of nonconformity.

Futile and superfluous elements become fundamental, and standards for what is considered beautiful and aesthetic have become highly important, thus maintaining the conditioning of unhappiness.

Regarding a world that is widely connected through social media, yet lacking the maturity to use it wisely and in a balanced way, a recent American study found that after about 15 minutes of reading gossip magazines or scrolling through the time line on Instagram or Facebook, there is a significant increase in cortisol levels and a reduction in endorphins, leading to a decrease in well-being, keeping the individual in a state of discontent.

In this context, the lack of idealism that transcends the immediate, the difficulty in dealing with frustrations, and immaturity in the face of everyday difficulties lead people to various physical and mental illnesses. Added to this picture of moral laxity is a dangerous ingredient: the vulgarization of sexuality at all levels of society, which is more dramatic for younger generations, anticipating and intensifying the natural blossoming of their sexuality when faced with countless appeals to sensuality present in all types of media (advertisements, music, dance, films, soap operas, and social networks).

According to Dr. Paulo Batistuta, this is a crisis of values with the loss of important references, at the expense of a superficial life and liberal discourse in postmodern society, such as the predominance of "*having* over *being,*" pleasure for pleasure's sake, and the permissiveness that justifies "everything is OK as long as I feel good." Today, there is a prevalence of relativism, where there is nothing absolute and truths are in flux; consumerism, where people live to acquire products and not values; and nihilism, characterized by subjectivity and a passion for nothingness, with frightening indifference.

We live in an age of intense competition and little solidarity, where the objectification of life and the predominance of personal interests over the collective are characteristic. Self-centeredness, superficiality and utilitarianism override ethics based on virtues, responsibility and respect for others.

In healthcare, the prevailing materialistic understanding in today's world holds that life is limited to the organic nature of the nervous system. This can lead professionals to act based on such an understanding, reducing the human being to a biological machine and justifying abortion, for example, before the nervous system is fully formed.

The media plays an important role in this context. Generally, media space is dominated by pro-abortion activists, many of whom are funded by NGOs, some of which are known to be involved in the fetal organ market in the U.S., creating a conflict of interest.

In fact, the debate surrounding abortion is clouded by ideological and economic issues that often obscure important aspects of the topic. Medical literature itself is influenced by social policies, and many articles are rejected by high-impact journals, while others are sponsored because of their ideological leanings.

The movements are structured mainly around the Judiciary (where the process is ultimately legalized). They usually employ so-called "trench tactics," seeking an extreme, controversial situation with media coverage in order to subsequently expand the progression of exclusions of illegality until complete legalization is achieved.

Thus, it is important to emphasize that the strategy used by pro-abortion movements is the desensitization of human life, simply removing the child from the discussion in the gestational process, and focusing discussions on the figure of the woman instead. There are pro-abortion structures, such as universities and the media, that make the unacceptable tolerable, then understandable, and finally an "indisputable right" and even a "necessity" for women.

Even in the face of this materialistic contemporary world view, it is truly naive to think that abortion frees women from a problem. The emotional consequences of someone who embraces a life that has been conceived are completely different from those of someone who chooses death. Carrying the weight of an abortion is completely different from sharing in the birth and development of a child, along with all the emotional bonds that follow in the various stages of life. This applies to the father as well.

The fantasy is created that, by means of abortion, "the problem" ceases to exist. As discussed in the previous chapter, several psychiatric studies show that women (rich or poor) who have abortions suffer some damage to their mental health.

The variables range from hormonal imbalance, caused by the artificial termination of the pregnancy, to emotional guilt, coupled with feelings of abandonment from unresolved grief. Thus, depression, anxiety, insomnia, and even psychosis are part of what some scholars understand as post-abortion syndrome.

It is understood, therefore, that the materialistic paradigm often distances society from a moral and ethical path, immersing individuals in a world of illusions, without noble purposes, without resistance to disappointment, and with little resilience for difficult times. Hedonistic concepts, which lead human beings to sensory exhaustion in pursuit of childish and immature pleasure, are encouraged by an industry that glorifies sexuality and consumerism to the detriment of family values and togetherness.

Therefore, a paradigm shift is urgently needed for humanity to mature properly when we grasp the fact that we are not just matter; that there is an immense universe beyond our feeble senses; that we need to take the non-material or spiritual plane into consideration and understand that the brain responds to the command of consciousness (of the spirit) and that we are not just a by-product of the electrochemical connections of neurons; that we are, in fact, living a learning and growth experience that did not begin in this lifetime and will not end at the completion of our current bodily existence.

In a society with little space and almost zero dedication to connecting with spirituality, where religions, immersed in dogmas and rituals, and more concerned with maintaining the status quo, often fail to establish a real connection between the individual and the Divine, science, with its investigations in various areas, is already beginning to play the role of true "priest of spirituality," and points to a beautiful path that will soon awaken humanity from its deep sleep in individualism and primitivism.

In the next chapter, we will present some new perspectives on science that definitively bring it closer to the non-material realm, filling us with hope for the future of humanity.

CHAPTER VI

SCIENCE AND SPIRITUALITY: A POSSIBLE AND NECESSARY INTEGRATION

"All truth passes through three stages:
First, it is ridiculed; second, it is violently opposed;
third, it is accepted as self-evident."
ARTHUR SCHOPENHAUER.

In 2014, more than 300 renowned scientists from various academic centers around the world, working in the areas of biology, neuroscience, psychology, medicine and psychiatry, signed a manifesto[14] for a post-materialist science. The event was organized by postdoctoral scholars Gary Schwartz and Mario Beauregard from the University of Arizona.[15]

Eighteen conclusions were highlighted by the scientists in favor of a post-materialist movement for science, a movement that takes into consideration the

[14] Mario Beauregard, Gary E. Schwartz, Lisa Miller, Larry Dossey, Alexander Moreira-Almeida, Marilyn Schlitz, Rupert Sheldrake, and Charles Tart. *Manifesto for a Post-Materialist Science.* Available online at: https://www.researchgate.net/publication/264463775_Manifesto_for_a_Post-Materialist_Science. – Tr.

[15] Lisa Miller, from Columbia University and listed in the preceding footnote, also contributed to the organization process. – Tr.

existence of a non-material component, one that is not restricted to the limits of matter, in the clear understanding that science should not be contained in any dogmatic system.

We will now highlight four main resolutions proposed by these cutting-edge scientists:

– Studies known as "psi-phenomena," psychic phenomena, or extrasensory phenomena indicate that in certain cases we can receive meaningful information without using our ordinary senses, in a format that transcends the usual time-space continuum. In addition, some research in the psychic field shows that minds can interact in a non-local way (according to quantum physics), without mediation, and without the information being degraded by distance, for example. Scientific materialism is unable to explain such common phenomena.

– Conscious mental activity can be experienced during clinical death in cardiac arrest (in the so-called near-death experience), during which, without electrical activity, the physical brain could not be responsible for the reported experiences.

– Scientifically controlled laboratory experiments have documented that individuals considered to be mediums can in some cases obtain detailed information about deceased individuals, supporting the subsequent conclusion that the mind must exist separately from the brain.

– Materialistic theories fail to explain how the brain generates the mind, being unable to explain the empirical evidence reported in psychic phenomena. This failure shows us that the time has come to free scientific thought from materialistic ideology to broaden the concept of the natural world, embracing the post-materialistic paradigm.

A few decades ago, medicine discovered the importance of spirituality in promoting health when several studies pointed to a clear relationship between religiosity/spirituality and a reduction in the incidence of suicide, better adherence to treatment in general and even faster recovery in cases of serious illness. As a result, leading American and European universities now offer courses on health and spirituality in their schools of Medicine.

We will now discuss some of the main points of convergence between science and spirituality, with the so-called non-material universe, an approach that we believe is the path to an important paradigm shift for humanity, a shift that will promote significant maturity from a moral and spiritual perspective.

6.1 "Near-Death Experiences" (NDEs)

Near-death experiences are events in which, after cardiorespiratory arrest, patients are resuscitated and then describe richly detailed experiences that, in most cases, have a major impact on their lives and those of their family members.

To qualify as an NDE, it is essential that the patient presents an isoelectric electroencephalogram, i.e. one that is without electrical brain activity. This means that, after 10 to 30 seconds of cardiac arrest, the cortical part of the brain, the region responsible for our thoughts, memory and perceptions, is completely inactive and unable to produce or capture any type of content.

This condition refutes all theories presented to date that seek neuro-electrochemical explanations for such experiences, when, in fact, the answer lies outside the brain, being in this case an experience of the individual's consciousness or spirit.

This phenomenon has been studied for many years, and in recent decades, scientific research has intensified, resulting in a vast body of literature with significant revelations. The first researcher and scientist who decided to tackle the subject in an academic and serious manner was Dr. Raymond Moody Jr., who wrote the book *Life after Life* in 1975, containing 150 detailed accounts of NDEs.

Another reference in such research is Dr. Sam Parnia, from the University of Southampton, in the U.K. Parnia documented more than 2,000 patients who had had a near-death experience, and he proposed establishing highly reliable comparison parameters for analyzing the cases, taking into account the drugs used during the patient's coma and resuscitation, the time interval without brain activity, and other clinical and laboratory parameters.

Reports of NDEs are becoming more and more frequent due to advances in cardiopulmonary resuscitation techniques and life-saving measures in hospital intensive care units; we now know that the chance of surviving a cardiac arrest is significantly greater than in the not-so-distant past.

Despite their individual nature, where each person recounts his or her own particular experience, near-death experiences have common elements documented by researchers, such as: out-of-body experiences, tunnels of light, contact with spirit beings (e.g., deceased family members and/or friends), recapitulation (as if watching a movie of all the important moments of one's life), and barriers (limitations on passage to the other plane).

A classic example is that of Dr. Eben Alexander III, a renowned neurosurgeon from Harvard, with a career spanning more than 25 years. In 2008, at the age of 54, he woke up one day with a severe headache. He was immediately admitted to the hospital and soon fell into a coma with a diagnosis of severe bacterial meningitis.

Dr. Alexander spent seven days in a coma, without neocortical brain activity (responsible for our ability to think, move, speak, hear, etc.), yet he had an intense experience that completely changed his life. At the time, he reported clear contact with another dimension of life, wherein he discovered places, music and people, including a biological sister whom he had never met.

The neurosurgeon spent four years meticulously analyzing the possible neurophysiological explanations for the rich and impactful experience he had had. In 2012 he decided to write about what he had experienced and tell the world that there is life after death, putting his credibility as a doctor at risk. He recounts his experience in detail in the book *Proof of Heaven*.

Another well-known example is that of American singer and songwriter Pam Reynolds Lowery, when at the age of 35 (1991), she experienced difficulty moving and frequent dizziness. Imaging tests revealed a large aneurysm near the brain stem (a very delicate area for surgical access). She underwent an unprecedented procedure at the time – neurosurgery with hypothermic cardiac arrest (a situation that left her without electrical brain activity).

After the surgery, the singer reported in detail a near-death experience, where she initially found herself outside of her physical body, floating in the operating room and watching the doctors operate. She felt completely conscious and was able to follow several surgical steps and even the conversations between the surgeons, a fact that was confirmed by the footage in the room, and which would not have been possible from a physical point of view, since she was completely without brain activity during the surgery.

Lowery claims to have been led by an invisible force to a large point of light, where she was warmly welcomed by deceased family members, including her uncle. She says the sensation was so pleasant that she did not want to leave, and her uncle had to lead her back to her body.

Dutch physician Dr. Pim Van Lommel, in a paper published in *The Lancet*, identifies permanent changes observed in patients after near-death experiences. His research followed patients and their families for eight years, finding significant changes in their outlook on life and worldview, such as:

- Reduced fear of death and material attachment
- Family appreciation
- Belief in life after death/spiritual interest
- Increased empathy and displays of affection

Thus, in scientific terms, we can reach an important neurophysiological conclusion: if mental activity exists without brain activity, the mind is not a product of the brain, but something that manifests itself through it.

6.2 Memory Regression (Past Life Therapy – PLT)

Another situation in which we see science approaching spirituality is in a field of psychotherapy that studies memory regressions, wherein experiences lived in previous existences are identified as the key to certain psychological conflicts that interfere with one's current quality of life. These experiences involve remote events that generate negative impulses or intense fears in the present, without correlation with any trauma in the present life that would justify such symptoms.

Thus, an important segment of the science that studies the human psyche has been increasingly focusing on past life regression therapy as an effective tool for bringing to light the individual's consciousness content that would otherwise remain buried in his or her subconscious, thereby enabling the reframing and disconnection of such traumatic experiences from the distant past.

Some cases involving specific phobias and symptoms similar to post-traumatic stress disorders have no explanation in the individual's current life history; rather, they derive from traumatic experiences from previous lives, emerging in the present as negative impulses or intense fears, for example.

In 1982, American researcher and PhD. in Psychology, Dr. Helen Wambach, published a study with approximately 750 cases of hypnotic regression. Her research used a statistical evaluation of past life regressions as a way to obtain an overview of such experiences to arrive at evidence of the reality of the phenomenon.

Renowned psychotherapists and psychiatrists have shared their positive experiences in articles and books. One such individual is the eminent American physician Dr. Brian Weiss, one of the most respected psychiatrists in the U.S., who revealed in his book *Many Lives, Many Masters* his first contact with past life regression, an event that was decisive in completely changing the paradigm of his clinical vision.

In his book, Weiss discusses the treatment he provided to a patient when, using a method of memory regression for early childhood, he ended up accessing a previous incarnational experience and thereby reached the core of his patient's trauma, which was decisive for the progress of her treatment.

Countering the argument that regressions are merely the imaginative creations of patients, there is scientific evidence that individuals undergoing past life therapy exhibit intense autonomic and neuroendocrine responses during the process, that is, there is an organic response to a real emotion when accessing a trauma or a significant moment in the distant past, a fact that would not occur if the situation were merely a figment of the imagination.

Corroborating this fact, Dr. Regis Ferreira, in his doctoral thesis, showed the variation of elements such as cortisol and adrenaline in patients undergoing memory regression therapy, indicating access to significant psychological stress, ruling out the hypothesis that the origin is substantially imaginative in PLT.

In another very interesting scientific study, authors Julio Peres and Andrew Newberg subjected some patients, during past life therapy, to functional neuroimaging (SPECT) brain scans, revealing the activation of areas linked to emotion and memory (cingulate gyrus and temporal lobe) and a lack of significant activity in the area responsible for creativity and imagination (frontal lobe), thus indicating that this is something linked to these patients' vivid memories and not their imaginative creativity.

Thus, part of science is beginning to accept what, according to Spiritist philosophy, would be the master key to accessing humanity's deepest reflections: reincarnation. Understanding it would enable us to know where we have come from, why we are here, and where we are going, based on our current choices.

6.3 Reincarnation

Conceptually represented by the spirit's return to the material experience, reincarnation is a fundamental pillar of the Spiritist Doctrine, with profound philosophical and moral consequences. Reincarnation has never been as close to academia as it is today, and the trend is for this natural law to be definitively incorporated in the not-too-distant future.

In the late 1970s, reincarnation moved from the religious/cultural sphere to the field of scientific research, including psychotherapeutic cases of past lives with verifiable details, near-death experiences, spontaneous memories of past lives in children, xenoglossia (knowledge of a language not learned in the present life), and studies of congenital marks associated with traumatic memories corresponding to previous lives.

Professor Hemendra Nath Banerjee was one of the pioneers in the scientific investigation of reincarnation, raising more than a thousand cases with a strong suggestion of reincarnation, when he wrote the book *Extracerebral Memory* in the 1960s, in which he reports discoveries of memories of past lives in his study.

Another pioneer of scientific research in this field was the eminent professor of psychiatry at the University of Virginia, Dr. Ian Stevenson, who, over the course of 30 years, studied thousands of spontaneous accounts from children regarding their most recent incarnations, documenting hundreds of relevant facts that the children could not have previously accessed. His extensive research produced numerous articles and books, such as the work entitled *20 Cases Suggestive of Reincarnation.*

Dr. Stevenson also organized a large file correlating birthmarks and defects, linked to the memory of traumatic events in a previous incarnation. He published the file in his two-part monograph *Reincarnation and Biology: A Contribution to the Etiology of Birthmarks and Birth Defects,*[16] which enables the reader to understand certain congenital malformations that have no plausible scientific explanation, that is, cases in which there are no correlations with genetic alterations or ingestion of teratogenic substances by the mothers, for example.

A very interesting case was studied by Dr. Erlendur Haraldsson, a professor at the University of Iceland. The case involved Purnima Ekanayake and the investigation took place between 1996 and 1999, with multiple interviews with numerous witnesses during several visits to the child's home country, Sri Lanka. Little Purnima lived in a small town called Bakamuna and, according to her parents, she had spoken of a supposed past life since she was 3 years old. She described a fatal accident with a large vehicle and that after the accident she said that she floated in the air in semi-darkness for a few days; she saw people weeping in mourning and her body at the funeral.

[16] A summary of Stevenson's findings are given in his book entitled *Where Reincarnation and Biology Intersect* (Westport, Connecticut: Praeger, 1997). – Tr.

Purnima said that her previous family had produced Ambika incense. At the age of four, while watching a television program about the Kelaniya temple (located over 300 km from her hometown), she stated that she knew about the temple and had lived on the other side of the Kelaniya River. She recounted specific facts about her life there, such as being an incense maker (providing details about how the product was made), having two wives (whom she named), as well as the names of her father, mother and siblings, and the school she attended as a youngster. Finally, when the data was confirmed by the researchers, something else was found: the accident that killed the former incense producer was a bicycle accident with a bus, which resulted in extensive injuries in the lateral thoracoabdominal region, documented in a medical autopsy, which coincided with birthmarks found on the child.

In Brazil, researcher Hernani Guimarães Andrade included in his book *Reincarnação no Brasil*, eight registered and duly proven cases of children who spontaneously remembered impressive details of their past lives.

Present since ancient times in the human psyche, we find reincarnation in the ancient books of the Vedas in Hinduism, in the doctrine of Buddha, in various religious denominations of African origin, in the interpretation of the Jewish Kabbalah, and in early Christianity, found in various passages of the New Testament, such as in the conversation with Nicodemus, when Jesus states: "No one can see the kingdom of God unless they are born again" (John 3:3), or when, after the Transfiguration, the disciples asked Jesus, "Why do the scribes say that Elijah must come first?" and Jesus replied, "Elijah has already come, as John the Baptist, and the scribes did not recognize him and did to him whatever they wanted..." (Matthew 17:10-13).

According to the Spiritist Doctrine, reincarnation is a fundamental law of nature and with it we can understand essential questions that have challenged human consciousness since ancient times:

- Where we have come from, where we are going and why we are on the earth.
- Material inequalities and all the supposed injustices of earthly life.
- The key to the interpretation of several passages in the Gospel of Christ.
- Morally and intellectually growth towards higher spiritual evolutionary stages.

The great humanist and medium Francisco "Chico" Xavier[17] once said that it would be difficult to conceive of the existence of God and Divine Justice without understanding reincarnation. Regarding Chico's statement, we propose the following reflections:

- Would there be divine justice in forever depriving happiness of those who committed wrongs during their existence? Let us suppose the belief in a single life for those who believe in God and that there is something after death; how could we understand the goodness of the Creator in ordering eternal punishment for those who had committed wrongs over the course of their lifetime? Let us imagine a mother who, after an existence aligned with moral and ethical principles, dedicated to family and society, would be eternally separated from her son (according to the conception of heaven and hell) because her son had lost his way in his earthly existence and took the path of crime, for example.
- Where would God's goodness and equity be in a fate determined after an existence filled with countless difficulties, or one of short duration?

Let us remember children afflicted with aggressive and incurable diseases. Would they then be singled out for such suffering in their "sole existence"? How can we understand that some are born into absolute poverty and barely obtain the food that sustains them, while others are born with "silver spoons," wanting for nothing materially?

Thus, reincarnation is the fundamental basis for explaining human difficulties, giving us the understanding of redemption from past wrongs via new experiences and comforting hope for those who have committed such wrongs. Therefore, no human being would be exempt from the process of education and evolution so that reincarnation would be the natural mechanism for this, without privileges or privileged individuals. The wording on Kardec's tomb, "to be born,

[17] Francisco Cândido Xavier (known as Chico Xavier) (1910-2002) was born in Pedro Leopoldo, Minas Gerais. He worked unceasingly as a medium and became known as the most devoted worker in the Spiritist Movement in the area of mediumistic activity, in addition to his devotion to the social and spiritual assistance and evolution of the human being. His psychographic production entailed more than 400 books, 87 of which have been published by the Brazilian Spiritist Federation. His "Life in the Spirit World" series, comprised of 13 books, are all available in English. – Tr.

to die, to be reborn again, to progress always, such is the law," perfectly sums up the meaning of reincarnation for Spiritism.

6.4 Mediumship

Mediumship, a natural condition that has historically been associated with mental illness and social maladjustment, has undergone significant conceptual change within society and even within the scientific community.

Initially, some psychiatric/psychological research on mediumship analyzed the issue as a classic dissociative identity disorder, a mental condition in which a single person displays characteristics that seem to belong to two or more personalities, each with unique ways and means of perceiving and interacting with the environment. This mental condition causes significant difficulties for the individual in his or her social environment, as well as obvious psychological distress.

Over time, other hypotheses were raised, and some authors pointed out the possibility that mediumship may actually be an expanded state of consciousness and not a personality disorder. In 2008, Dr. Alexander Moreira[18] published a scientific article in the *Journal of Nervous Mental Disease* entitled "Comparison of Brazilian Spiritist Mediumship and Dissociative Identity Disorder," comparing mediums with patients with dissociative identity disorder (DID), a mental disorder that can inevitably lead to serious cognitive and social impairments. According to Moreira's study, there are several factors that these two groups have in common, such as perceptions of voices or supposedly hallucinatory visions. However, unlike the pathological state of patients with dissociative disorder, the mediums who were studied presented better mental health and social adjustment than the average of the general population in psychiatric/psychological evaluation tests, concluding that mediumship is definitely not a dissociative mental disorder.

According to the Spiritist Doctrine, the ability to interact with the non-material plane is inherent to human beings; it is a biological property of sense perception, which is more highly developed in some people, although it does not endow them with privileges regarding morality, religiosity, education or social class.

---

[18] Alexander Moreira: Full Professor of Psychiatry at Federal University of Juiz de Fora, Minas Gerais, Brazil. Managing Director of NUPES – Center for Research on Spirituality and Health. In 2025, he was awarded the Oskar Pfister Award from the American Psychiatric Association. – Tr.

Throughout his medical career, Cambridge neuropsychiatrist Dr. Peter Fenwick observed many reports of spiritual perceptions in the final moments of his patients. He conducted a study involving 38 experienced professionals (including doctors and nurses) and 200 terminally ill patients and found that about 62% of patients or their companions reported spirit apparitions or visions of deceased family members on their deathbed, and that these perceptions usually had a positive impact on the professional team and the patients' families.

Another interesting study was conducted in 2007 at the University of Arizona, when a group of mediums was studied using a triple-blind scientific protocol to identify the possibility of receiving information about deceased individuals. The study concluded that some mediums in the study showed a significant ability to receive information about deceased individuals, ruling out the possibility of telepathy and confirming the data with family members. The study was published in the journal *Explore*, under the title "Anomalous Information Reception by Research Mediums Demonstrated Using a Novel Triple-blind Protocol."

Dr. Emily Williams Kelly of the University of Virginia's Department of Psychiatry conducted research with similar results in 2011, which were published in the *Journal of Nervous Mental Disease* in an article entitled "An Investigation of Mediums who Claim to Give Information about Deceased Persons."

Another highly significant and unprecedented study ("Neuroimaging during Trance State: a Contribution to the Study of Dissociation") was conducted at the University of Pennsylvania in 2012, when Dr. Julio Peres and his colleagues demonstrated that psychographic mediums activated different areas of the brain in functional neuroimaging tests after transcribing their own text, which was then compared to a psychographed text, showing that mediumistic texts come from another mental source.

Peres joined forces in 2017 with Moreira and Alessandra Mainiere to conduct research at Aachen University in Germany, using functional magnetic resonance imaging (fMRI) of the brain to demonstrate the difference in brain activation during the so-called mediumistic trance state. The study involved a group of eight mediums and a control group of eight individuals, all of whom had similar clinical neuropsychological assessments.

All the volunteers underwent functional magnetic resonance imaging while in a normal state, and then the mediums underwent a new sequence of exams, but in a state of mediumistic trance, displaying significant variations in the activation of the neurosensory and auditory systems at the time of trance in comparison to the resting state and in relation to the control group. The results were published under the title "Neural Correlates of Psychotic-like Experiences during State of Trance."

Given the above information regarding mediumship, it is extremely important to consider the existence of extra-physical perceptions in healthy individuals unaffected by mental disorders, since still-predominant materialistic science leans toward the diagnosis of psychotic disorders. In line with this need, a multicenter study, involving 52 countries and 250,000 participants, stands out, reporting psychotic symptoms in about 12.5% of the participants, with less than 1% diagnosed with schizophrenia (Nuevo et al., 2012); in other words, most individuals who reported seeing something that was not on the material plane had normal neuropsychological assessments. Hence, the mediumistic ability of certain individuals should be considered a real possibility for explaining these phenomena.

Science is based on the impossibility of certainty. It is important to consider the provisional aspect of science as a necessary element for its own evolution, remembering that what we label as "myths" today are ancient accepted truths, just as what is now considered "unrealistic fiction" may become future scientific reality, a time in which we identify humanity's surrender to the inevitable path of progress.

CHAPTER VII
THE SPIRITIST DOCTRINE

"To be born, to die, to be born again and to progress: such is the law."
ALLEN KARDEC

The dogmatic religious system that imprisoned science and philosophy in the Middle Ages suffered a major rupture with the advent of the Enlightenment in the 18th century, a movement that became fundamental to the intellectual growth of humanity and allowed the development of critical thinking and the debate of ideas.

That historic era was fundamental for the improvement of scientific methodologies and for the carrying out of major research, leading to discoveries in the field of science, unlike anything that had ever been seen in history.

From the end of the 19th century onward, the evolution of humanity's scientific thought called human beings to enter a new era, marked by a period of phenomenal scientific/cultural growth that would change their understanding of the macrocosm, the microcosm and everything in between; it would represent an important step toward the future understanding of the Divine laws, including immortality as one of the fundamental attributes of the soul.

Thus, the right moment was established for "an orchestrated, timely invasion" originating from the spirit plane, when, in the second half of the 19th century, the spirit world began to manifest itself in various parts of the planet, based on so-called "physical effects phenomena," i.e. events that would occur in a short space of time, with the clear purpose of awakening humanity to non-material life, or life on the spirit plane.

A notable event of this historical moment was the phenomenon that occurred in Hydesville, New York in the Fox family home, with the manifestation through physical sound effects of the spirit of the murdered peddler Charles Rosma, culminating in the discovery of his corpse.

At the same time, physical phenomena began to occur throughout Europe, stimulating research by the scholar Professor Leon Hipolites Rivail, a disciple of Pestallozi. Using a rigorous methodology through which he initially sought to demystify paranormal phenomena, Rivail, under the pseudonym Allan Kardec, would eventually become largely responsible for codifying the greatest and most profound Spiritualist treatise in our history. Uniting science, philosophy and religion, the Spiritist Doctrine emerged.

A controlled and silent revolution began with the publication of *The Spirits' Book*, and a gradual change in human consciousness and conceptions began to take place.

Spiritism emerged as a consoling doctrine that came to redeem – in the words and example of Christ – the figure of God with a new look, combining philosophy and religion with scientific achievements to understand the greatness of a God of infinite love and mercy, showing us the profound meaning of our existence and the process of our ongoing evolution.

Since the Spirit World respects the sequence of human progress, resulting from our own efforts, the foundations of Spiritist thinking were postulated at a precise historical moment for their understanding, with the maturing of scientific thought that would come especially at the beginning of the 20th century, based on Einstein's work and discoveries in quantum physics.

The methodology for constructing the entire Kardecist "Pentateuch" (the five books that form the basis of the doctrine) was similar to that of the positive sciences: the experimental method.

According to this method, new phenomena, which arise and cannot be explained by known laws, are observed, compared and analyzed, tracing effects back to causes to arrive at the governing law. Then, consequences are deduced and useful applications sought. It was not the phenomena that came a posteriori to confirm the theory; it was the theory, which subsequently explained and summarized the phenomena.

The clarity of Spiritist explanations reveals to us the incessant evolution and progress of human beings, a fact that occurs by means of the individual effort of their intelligence, along with the help of those who are more advanced serving as veritable tutors, in addition to the presence of certain individuals involved in missionary activities serving as examples.

For Spiritism, the greatest example, model, and guide for humanity is Christ. The Spirits reported that Jesus had been appointed by God to coordinate the terrestrial sphere from the very beginning of its formation four billion years ago.

"I have many things to tell you that you cannot yet understand, and I have many more to say to you that you will not understand. Therefore, I speak to you in parables. But later I will send you the Comforter, the Spirit of Truth, who will restore all things and explain all things to you." (John 14; Matthew 17).

Spiritism, based on Christ's own words, just as Christ based his teachings on those of Moses, is a direct consequence of His doctrine. Through Spiritism, human beings know where they have come from, where they are going, why they are on this earth and why they suffer temporarily; they can also see God's justice everywhere. They understand that, in general, all suffering is the result of unfortunate actions in the past, and is always a great opportunity for spiritual growth, depending on how it is managed.

Spiritism also explains that souls progress incessantly, through a series of successive existences, until, more or less quickly, and through work and good will, they reach a level of perfection that brings them closer to God.

It teaches that there are no disinherited individuals, nor any that are forever doomed to evil and suffering; that those who are called demons are actually spirits who are still backward and imperfect, who practice evil in the spirit realm just as they did on the earth, but who will advance and improve in their own time; that angels or pure spirits are not separate beings in creation, but spirits who have reached their goal after having arduously traveled the road of progress.

Kardec was responsible for the largest compendium on mediumship ever written, and *The Mediums' Book*, specifically, is an important guide for the safe exercise of that sense perception-based biological faculty, present in an ostensible way in a significant part of the population.

Through the exercise of mediumship, aligned with the moral precepts exemplified by Christ, Spiritism was able to lift the veil of misunderstanding about the spirit world, offering the comforting reality of the immortality of the soul.

A wonderful example of the reach of a Christian mediumistic practice was exemplified by the life and work of Francisco Cândido Xavier, who psychographed over four hundred books and countless letters of consolation. These so-called consolatory letters soothed the hearts of thousands of mothers, fathers and family members, restoring hope as they understood that separation from their loved ones is temporary and that, fundamentally, life goes on.

In 2014, Dr. Moreira and his collaborators, using rigorous methodological analysis, published a scientific paper on the content of the letters psychographed by "Chico." Their conclusion was that the letters conveyed precise and accurate information, dismissing the most diverse explanations from a materialist point of view.

Spiritism introduced the concept of reincarnation more broadly into the Western world, revealing that all spirits go through multiple incarnations as part of a continuous process of progress and evolution. The doctrine of reincarnation is opposed to the idea of free salvation through faith; it values the individual effort of the immortal spirit, which gradually improves emotionally and intellectually over millennia. This journey represents a manifestation of the divine inheritance, granted equally to all spirits. Individual differences are not the result of divine whim, but rather of each spirit's dedication to promoting its own progress.

Over the course of its journey, each spirit reaps the fruits of the good it has sowed and it is also compelled by justice to make amends for the wrongs it has committed. In this way, divine mercy softens and offers different opportunities for growth to the erring spirit, helping it overcome the weight of guilt on its conscience. What some religions see as punishment is, in fact, for Spiritism, the natural consequence of past wrongs.

By studying reincarnation, Spiritism demonstrates that religious truth is not incompatible with scientific truth. Spiritual evolution occurs gradually, simultaneously with the physical evolution demonstrated by Darwin. Furthermore, Spiritism restores, in human consciousness, one of the essential attributes of a Perfect Being: justice.

Reincarnation is, therefore, the key to understanding the great philosophical questions of humanity, by handing over the weight of responsibility and the credit of meritocracy to the individual, who thus understands the proposal of learning (whether through trials or expiations) from every difficulty that reaches him or her. The harshness of social injustices and their structural anomalies, as well as great pains and individual losses, are reinterpreted in the certainty that, governed by just, Divine laws, and under the tutelage of a merciful and good Creator, we are touched by the difficulties we ourselves create; the Divinity allows us to experience vicissitudes that we are capable of enduring, and facing them is part of our invitation to maturity.

Chico Xavier once said that, without understanding reincarnation, it would be very difficult to conceive of the existence of God; for example, how can one explain to parents why their child has died? Or why, with all the potential for life, does a young person develop aggressive cancer? Without understanding the multiple existences in this eternal school of life, it would be difficult to glimpse meaning in so much pain.

Experimental Spiritism studied the properties of the spiritual fluids and their action on matter. It demonstrated the existence of the perispirit, suspected since ancient times and designated by St. Paul under the name "spiritual body." Corresponding to an inseparable "envelope" of the soul, one of the constituent elements of the human being, the perispirit is the vehicle for the transmission of thought and, during the life of the body, it serves as a link between the spirit and matter.

An important feature of the Spiritist revelation is that, based on facts, it must be essentially progressive, like all observable sciences. It is allied with science, which, being the exposition of the laws of nature, cannot be contrary to the laws of God, the author of those laws. According to Kardec, if at any point science contradicts Spiritism, science must always prevail.

Therefore, in the preface to the book *In the Realms of Mediumship* by André Luiz, psychographed by Chico Xavier, the mentor Emmanuel states:

> Raised to the status of investigators of the truth, chemists and physicists, geometricians and mathematicians have become today's priests of the spirit without desiring it. Due to their ongoing studies, materialism and atheism will be compelled to vanish for lack of matter, the base that had ensured their negativistic speculations.[19]

[19] Francisco Candido Xavier, by the Spirit André Luiz, *In the Realms of Mediumship* (Brasilia: Brazilian Spiritist Federation, 2011), 8. – Tr.

CHAPTER VIII

HUMAN LIFE IS NOT AN ACCIDENT

"Verily I say to you: None can see the kingdom unless they are born again."
JOHN 3:3

Spiritism lifts the veil of misunderstanding about the spirit world. This comforting doctrine provides us a completely rational way of understanding the immortality of the soul, giving profound meaning to our existence and the process of continuous evolution in which we are immersed. By means of this understanding, we can comprehend where we have come from, why we suffer, and where we will go after our experience in matter is over.

The knowledge of reincarnation as a natural law enables us to understand Divine justice and mercy; each return to life (reincarnation) brings the abilities of the previous planting (in the form of our innate qualities and talents), but also the weight of the wrongs we committed, making us responsible for our own choices.

Each corporeal experience is structured for the spirit's progress; it is a meticulous attempt to welcome, among friends and enemies, the path of reunion with our neighbor, added to the possibilities of learning and the intellectual and moral development that each opportunity offers.

According to Spiritist science, the spiritual body (called the perispirit) is closely linked to the entire gestational process. For example, it is responsible for inducing a specific sperm to meet the female egg in the distal third of the fallopian tube, with both being selected according to the physical needs of that particular individual. This would explain why the winner of that grand marathon, which lasts approximately seven hours from the time it enters the uterine cervix, then undergoes the selective process of capacitation in the fallopian tubes, and finally completes the fertilization process (union of male and female gametes), is often a sperm with structural alterations.

Even more impressive is the fact that, to this day, embryology does not know what coordinates the process of cell differentiation in the embryo, a mechanism in which identical cells differentiate meticulously and transform into completely different cells, thus giving rise to the various tissues and organs of the human body. It used to be believed that the DNA itself managed this process, but genetics failed to determine any specific gene sequence that could coordinate the action. In reality, according to High Order Spirits, the great conductor of this journey is the reincarnating individual's spiritual body (perispirit), which is connected to the embryo at conception, acting to guide and modulate the entire process of cellular differentiation and organization.

When academic science breaks through the walls of materialism, this possibility will inevitably be considered.

In this context, Spiritism sheds light on so-called biological "flukes," when, contrary to expectations, a baby is born, for example, with some type of disability, a situation that is usually very difficult for those involved. It is common, on such occasions, to question divine justice and goodness when a child is "randomly condemned" to a defective body. Only the law of reincarnation can sufficiently elucidate the intricacies of these abnormalities caused by changes in the number or structure of chromosomes, individual genetic mutations, or congenital malformations.

According to Spiritist understanding, these congenital alterations may be part of a healing process for the spiritual body, which purifies itself by purging the physical body of its ailments. Or, it may even be a choice made by the spirit itself as a means of trial for its personal growth. It is also understood that there is no such thing as chance when it comes to family constitution; one is a member of the family nucleus destined for his or her process of redemption and maturation.

In these cases, the parents are naturally subject to the Law of Causality and are redeeming their criminal complicity, or they are lovingly involved in the maturation of the spirit that incarnates as their child.

In response to cases of special-needs children, Chico Xavier's Spirit mentor, Emmanuel, states: "Motherhood is a privilege that God has granted to women. Every woman enjoys this privilege of Divine Providence, but special-needs children are entrusted only to exceptional women, who demonstrate an infinite capacity for love."

Thus, the Spiritist Doctrine always positions itself in favor of life and against abortion, with one exception, as per the following passages from The Spirits' Book:

358. *Is artificial abortion a crime regardless of the time of conception?*

"A crime is always committed when the law of God is transgressed. The mother or any other person involved always commits a crime upon taking the life of a child before its birth because it prevents a soul from experiencing the trials for which the unborn's body was to have been the instrument."

359. *In cases where the life of the mother would be endangered by the birth of the child, is it a crime to sacrifice the child in order to save the mother?*

"It is better to sacrifice the being who does not yet exist than the being who already exists."

Spiritism reinforces the fact that the right to life is a natural right. Furthermore, it explains that it is the first natural right of humankind. That is why no one has the right to attempt to take the life of another, or do anything that could compromise his or her physical existence. Life is an inviolable right.

8.1 Eugenics

The practice known as eugenic abortion, which is performed due to fetal anomalies, is also considered a crime, according to the Spiritist doctrine, because in such cases we are violating the laws that govern life. No one has the right to attempt to transgress these laws, for life is an inexorable imperative that cannot be cut short when it comes to the physical body.

Spiritism has a profound scientific perspective, which reveals aspects previously incomprehensible to human investigation.

In most cases, anencephalic babies are spirits who have been reborn without part of their central nervous system as a result of having attempted to take their own lives in a previous existence, either via suicide or an injury that affected their spiritual body (perispirit - the biological organizing model that guides the formation of the physical body during the reincarnation process). They return to physical life to repay their debt in a very brief existence.

We are not allowed to take a life under any circumstance. Anencephalic babies have the right to live. There are examples of anencephalic babies who lived for a long time and were loved by their parents until they fulfilled their designated purpose, which was to complete the stage of life that had been prematurely cut short. In this way, they naturally discarnate and reestablish themselves in the spirit world. Often, they are reborn into the same family as before, providing their loving parents with the happiness they could not previously enjoy.

Similarly, mothers who face this extreme pain are usually also involved in the redemption of these reincarnating spirits, because they were often jointly responsible for the criminal act in their previous life or, because they loved them so deeply, they committed to being their mother so that they could achieve the "purification" of their spiritual body. Although it entails great anguish, this experience is part of the Divine Plan, a necessity for future mothers, who are deserving of support and solidarity to alleviate their afflictions.

Consequently, it is crucial that such mothers understand and accept the fact that the children given to them by the Divine have congenital disabilities; that they wait a little longer and offer their love so that their children can fulfill their spiritual commitment. If they are stillborn or if they die shortly after birth, in either case, they will be completing their learning journey, while their mothers will have played a role of great elevation before their own consciences and the Sovereign Codes of Life.

Thus, any legal justification for taking a life will always be an ethical transgression, as human justice often finds itself at odds with Divine Justice. In 1935, when Adolf Hitler came to power in Nazi Germany, he began his reign by making abortion legal. On the tragic Night of Broken Glass, he authorized the mass murder of millions of people. Soon thereafter, in the same wave of violence and delirium that consumed him, the German leader urged parents to send their children with disabilities, including those who suffered only from bedwetting, to the gas chambers and extermination camps in the name of "purifying" the German race. Thousands of children were killed in the name of a supposed superior race that existed only in the sick imagination of a psychopath. He then authorized the mass murder of more than six million Jews and more than three million people of other ethnicities, simply because he considered them inferior, exalting the Germanic ideal.

In the early 20th century, Stalin sent thousands of political enemies into exile in Siberia, as well as forcibly committing those who disagreed politically with state doctrine to psychiatric hospitals or executing them by lethal injection. In both Germany and the Soviet Union, thousands of lives were summarily wiped out with "legal" support and various justifications for these atrocities.

In effect, the legalization of eugenic abortion paves the way for bolder steps in the future, favoring the intolerable widespread practice of abortion.

8.2 Rape

Regarding rape, one of the cruelest and most revolting acts of violence committed by men, the Spiritist position continues to be absolutely opposed to abortion, even though it is supported by human legislation.

It is understood that, in the case of rape, when the woman does not feel psychologically prepared to raise the child, it would be up to society and government agencies to facilitate and encourage the adoption of the child, instead of promoting its legal death.

In fact, the infant's right to life should not be either less important than, nor contrary to caring for the psychological well-being of the abused woman, especially because, as already discussed in previous chapters, the psychological and emotional consequences of abortion are also devastating for the mother's health.

Furthermore, in these cases, would it be just to ask if it was the infant that committed the crime? The answer is obviously no, so why hold it responsible for a crime in which it had no part?

If Brazilian law does not even provide for the death penalty for the criminal who committed the heinous act, how then could we condemn to death a defenseless fetus, with absolutely no guilt in the violence committed?

8.3 Spirit Obsession

Decades before science reached conclusions about the increase in the rates of various pathologies in women who had chosen abortion, André Luiz, in a book psychographed by Chico Xavier, in 1957, stated that the woman who opts for it is forced, by irrevocable laws, to suffer depressing changes in the genetic center of her soul, generally predisposing herself to painful illnesses, such as vaginismus, uterine infarction, cancerous tumors and scourges, often leading to her discarnation...[20]

The dire consequences of induced abortion do not end there, however. Spiritually, the repercussions of parents' criminal irresponsibility for rejecting those who are to return to the flesh – those with whom they had often made sacred commitments – are even more delicate.

Feeling robbed or betrayed, many of these aborted entities begin to harbor a deep hatred toward those who refused to welcome them into a new cradle and they make their earthly existence a living hell, subtly accompanying them until they experience some difficulty in life. At that moment, they exacerbate their existential conflicts until, in long obsessive processes, they manage to throw them into depressive states and various organic dysfunctions, or they await them in the spirit realm, thirsty for revenge, fostering apparent psychotic conditions, which are in fact mediumistic disorders, classified as spirit obsessions.

In some cases, even when it is not yet aware of what has happened to it, the spirit of the fetus is magnetized by affinity to the one who expelled it from the womb, which in the future may cause cervical cancer and other diseases of the female reproductive system. The hatred of these beings whose lives have been curtailed turns into a madness that is converted into harmful mental waves in the woman's weakened body, which assimilates them and causes changes in her state of health, thus characterizing a process of spirit obsession.

8.4 Redemption

In any situation in human life, it is important to forgive oneself for one's wrongs, and this opportunity extends especially to those who have had an abortion. Self-forgiveness is a fundamental milestone for proper psychological and spiritual healing. The Spiritist instructor Divaldo Franco explains that, just

---

[20] The actual quote may be found in *Action and Reaction* by Francisco Candido Xavier, dictated by the Spirit André Luiz (Series: **Life in the Spirit World,** International Spiritist Council, 2010),), 207: "To snatch a child from the maternal womb is confessed infanticide. The woman who commits or whitewashes such a crime is constrained by irrevocable laws to suffer dreadful anomalies in the reproductive center of her soul's perispiritual envelope. She predisposes herself to dolorous infirmities such as metritis, vaginismus, metralgia, uterine infarction and uterine cancer, from which she will very often discarnate..." – Tr.

as forgiving others does not mean forgetting the offense suffered, self-forgiveness does not involve erasing from our memory the wrongs we have committed. In this sense, just as forgiving others does not require unconditional trust in those who have hurt us, self-forgiveness does not allow us to escape responsibility for our actions. Furthermore, in order to forgive ourselves, we must abandon self-punishment, which is harmful to our mental health.

The Spiritist Doctrine is based on Christ's lessons of love, and it holds that there is no wrong that cannot be righted. All of us, without exception, commit wrongs, and when this happens, the Higher Laws afford us, at any time, the opportunity to rehabilitate ourselves and repair them. The only being who came to earth without committing wrongs was Jesus, for he came as a perfect Spirit.

There are various factors in our youth, maturity and old age that influence our misguided actions, whether consciously or unconsciously. Therefore, for those who commit abortion, in addition to aggravating factors, there are also mitigating factors, such as ignorance, adverse circumstances that led to the act, social pressure, and psychological conflicts.

These factors alone represent the beginning of rehabilitation for the crime; all of this mitigates the guilt. However, the most significant mitigating factor is the good deeds we perform, since we are not on the earth to pay for, but rather to redeem, through love, the wrongs we have committed. Committing wrongs is common, and rehabilitating ourselves is a moral duty.

Thus, whoever committed the wrong of having an abortion should forgive herself. Free herself from guilt! And then love! Divine Law is one of tenderness and compassion, not of heartlessness, since God is infinitely loving.

When we commit a wrong, we disrupt the Universal Order; therefore, our rehabilitation occurs when we restore that same order. All the good deeds we perform will be recorded to compensate for the wrong.

Even if the aborted spirit is still in the spirit realm, if it is emotionally connected to the mother, it can be reborn and reach her arms (not necessarily as her child), because the Law is inexorable and always brings the "debtor closer to the debt collector."

Drawing on Spiritist principles, Divaldo Franco advises women who have opted for abortion:

> Adopt a child affectively; it is not necessary to take him or her home. Become his or her benefactor, helping the family with major material

needs. When a child needs education, assistance or medicine, help him or her in honor of your unborn child and be at peace with your conscience. Or support a suffering person. Practice charity. Each gesture of love will gradually eliminate the wrong you did.

Help your fellow human beings in some way, and let love take care of lighting up the dark landscapes of your crime, for "love covers a multitude of sins."

Unfortunately, it is always the woman who bears the worst burden. We always talk about the one who had an abortion, but we do not talk about the one who induced her to have it, be it the irresponsible father, the professional or non-professional person who performed the surgery to criminally terminate the pregnancy.

Sometimes, the young woman indulges in an affective-sexual relationship out of inexperience. And when she finds herself ruined, abandoned and unhappy, she ends up opting for the worst because of several factors. This happens due to the stimulus of a materialist culture and certain bizarre behaviors of people who have become famous. But one day, this young woman's conscience will tell her that she has chosen a crooked path.

Carl Gustav Jung stated that, in his clinic dedicated to the treatment of depression, the greatest number of women over the age of forty who had a depressive condition had had abortions in their youth. When they reached the age of reason, they realized the crime they had committed when they were inexperienced.

Do not cultivate a conscience of guilt and do not look back! Do not punish yourself and do not give in to depression! Via reincarnation, the Divinity gives us the blessed opportunity to grow and love.[21]

[21] Divaldo Franco, *Sexo e Consciência*, (Salvador: LEAL Ed., 2013), 106-107. – Tr.

CHAPTER IX
LIFE IS WORTH LIVING

"I shall take the purity of the children's response:
And life, it is beautiful and it is beautiful..."
GONZAGUINHA.

9.1 Life and its Philosophical Meanings

Far beyond physical and biological meanings, the understanding of what "Life" is necessarily involves a philosophical and spiritual perspective.

Throughout history, the definition of the Human Being and Human Life has always been a subject of concern and study on the part of great philosophers.

Initially, the pre-Socratic philosophers were responsible for inaugurating philosophical thought as we know it today. They were the first to break with the exclusively mythical explanation – based on myths – that prevailed in Ancient Greece between the 7th and 5th centuries B.C.

Until then, in Greece and other civilizations of the time, only mythology explained the origin of the universe and human life.

With the awakening of philosophical thought, people began to seek an understanding of Life that involved, at the same time, the notion of rationality and transcendence of being.

The philosopher Socrates (470–399 B.C.), famous for having initiated the Socratic or Anthropological Period in ancient Greek philosophy, was the first to engage in the study of essentially human issues arising from human coexistence in society, such as the "fair life," goodness, morality and truth. He earnestly argued that any further knowledge of the world would require, above all, *a constant search for self-knowledge*, and he was largely responsible for the phrase *"I know that I know nothing,"* expressing the maxim of humility and the search for self-knowledge until the end of his days.

Later, Plato (428-348 B.C.), Socrates' main disciple, continued his master's ideas, developing his theory on the "Immortality of the Soul" or "One Life." According to Plato, the body (transitory) would be the vehicle of the soul, which in turn would be permanent and eternal; in other words, the soul would pre-exist the body, exist in the body, and continue to exist after the extinction of the body. In his work *Phaedo*, he offers numerous teachings about human life and the need for humans to live the present moment with greater awareness and intensity.

With Plato, we understand that life is a constant search for knowledge and truth, a journey to attain wisdom and transcend the physical and transitory reality of the tangible world. This philosopher viewed life as a unique opportunity to purify the soul and free oneself from the "cycle of reincarnation," thus reaching the world of eternal and immutable forms. In this way, the ideal life would not be that of the apparent world—the visible world—but rather one in which an individual constantly seeks knowledge and wisdom through dialogue and inner reflection.

The same philosophical movement can be observed in the area of Stoic thought, whose great exponents were Seneca (4 B.C. – 65 A.D.), Epictetus (50 – 138 A.D.) and Marcus Aurelius (120 – 180 A.D.), philosophers whose lessons about life remain practical and current many centuries after their emergence.

These thinkers, even before Christ, taught that life was the most important natural and spiritual phenomenon to be understood and studied by humanity. According to Epictetus, all events that befall humanity have their origin in nature, and nothing is outside a "Law of Universal Meaning," so that the quality of things, whether "good or bad," depends exclusively on our way of thinking about them.

In other words, the Stoics held that the human mind would be the great vehicle for generating either peace or anguish in our existence: when faced with a difficult situation, it would be up to humans individually to decide whether to face it as something natural on the path and conducive to some learning, or to victimize themselves in the face of life, as if the universe were conspiring against their personal happiness.

The stoic and noteworthy Roman emperor Marcus Aurelius stated:

> That which rules within [the soul], when it is according to nature, is so affected with respect to the events which happen, that it always easily adapts itself to that which is and is presented to it. For it requires no definite material, but it moves towards its purpose, under certain conditions, however; and it makes a material for itself out of that which opposes it, as fire lays hold of what falls into it...[22]

In this way, the Stoics offered valuable lessons on how to be more resilient in the face of life's events, how to transform one's mind into one's personal refuge, and how to learn to value the present as an opportunity to learn and achieve a more meaningful and rewarding life.

Thus, we understand that the study of Philosophy is a valuable instrument through which individuals can transcend worldly concerns and reach a state of wisdom and inner enlightenment, capable of providing feelings of peace, serenity and well-being in their daily lives.

Indeed, it is clear that since ancient times, the search for the meaning of Human Life has invariably passed through the notion of transcending matter, understanding the reality of the "Soul" and continuously searching for knowledge and self-improvement – what the Spiritist Doctrine would years later call "inner reform."

Many centuries after Plato, Chico Xavier, aligned with the pillars of Spiritism, taught that life would be a continuous process of learning and spiritual evolution; that understanding reincarnation would make it possible to achieve the purpose of evolving morally and intellectually with each birth, and in this way, the challenges and trials we face during our earthly journey would help us develop virtues and overcome imperfections.

9.2 Life is Worth Living

From all the perspectives studied so far – biological, scientific, spiritual and philosophical – human life is a precious gift that is worth experiencing.

---

[22] Marcus Aurelius, *Meditations* (Standard EBooks, translated by George Long), 28. Available online at: https://archive.org/details/marcus-aurelius_meditations_george-long/page/n25/mode/2up?q. – Tr.

The verb *To Live* is imbued with movement; it entails constant advancement and evolution from the first biological cell that forms, and it never ceases to expand, from the complete preparation of the embryo, to the fetus, to the infant, to the child, to the young person, to the adult, to the elderly person. Each of these phases involves complex physical, material, spiritual and philosophical processes, which comprise the integral constitution of the Being.

Living, according to the poet Khalil Gibran, is "having the vision of infinity and all the possibilities and achievements that love can bring us"; according to the musician Gonzaguinha,[23] "living is singing the beauty of being an eternal learner…"; according to Charlie Chaplin, "life can be wonderful if you're not afraid of it…"[24]

In the book *Man's Search for Meaning*, author and psychiatrist Viktor Frankl[25] states that "to live is to find meaning in pain." A former prisoner of the Auschwitz concentration camp, he recounts his experience and the extremely difficult situations he endured in that environment of grief and deprivation. Faced with such situations, he narrates how the survivors managed to react to all the evils they experienced and how they found meaning in life after this painful experience. In general, those who managed to survive or bravely endured their final moments in inhumane confinement were those who learned, in some way, to transform their personal tragedy into their triumph, and to convert suffering into a human achievement.

In the words of Frankl:

---

[23] Luiz Gonzaga do Nascimento, Jr. (1945-1991), aka Gonzaguinha, was a famous Brazilian singer and composer. – Tr.

[24] A line by Chaplin's character "Calvero" in the 1952 film *Limelight*. The full quote is, "Life can be wonderful if you're not afraid of it. All it takes is courage, imagination ... and a little dough." (https://www.goodreads.com/quotes/254535-life-can-be-wonderful-if-you-re-not-afraid-of-it). – Tr.

[25] Victor Frankl (born March 26, 1905, Vienna, Austria—died September 2, 1997, was an Austrian psychiatrist and psychotherapist who developed the psychological approach known as logotherapy, widely recognized as the "third school" of Viennese psychotherapy, after the "first school" of Sigmund Freud and the "second school" of Alfred Adler. The basis of Frankl's theory was that the primary motivation of an individual is the search for meaning in life and that the primary purpose of psychotherapy should be to help the individual find that meaning. https://www.britannica.com/biography/Viktor-Frankl. – Tr.

> When a man finds that it is his destiny to suffer, he will have to accept his suffering as his task; his single and unique task. He will have to acknowledge the fact that even in suffering he is unique and alone in the universe. No one can relieve him of his suffering or suffer in his place. It is one of the basic tenets of logotherapy that man's main concern is not to gain pleasure or to avoid pain but rather to see a meaning in his life. That is why man is even ready to suffer, on the condition, to be sure, that his suffering has a meaning.[26]

A similar idea was defended by Friedrich Nietzsche, when he said: "He who has a why to live can bear with almost any how."[27]

These authors help us understand that the beauty of life can be found in all situations, even the most adverse ones, as long as we seek to accept the notion that life is a learning experience and that pain, suffering and difficulties are teachers that force us to learn lessons along the way.

Furthermore, if it is true that the earthly experience will always bring us some degree of suffering and difficulty, it is also true that by means of it we have the opportunity to exercise the most beautiful sentiment that can be learned by human beings: love for one's neighbor.

Still drawing on the lessons of Viktor Frankl, he wrote that

> ... love is the ultimate and the highest goal to which man can aspire. Then I grasped the meaning of the greatest secret that human poetry and human thought and belief have to impart: The salvation of man is through love and in love. I understood how a man who has nothing left in this world still may know bliss, be it only for a brief moment, in the contemplation of his beloved.[28]

The experience of love in human relationships is so important that it is even being verified in scientific research. For over 80 years, researchers at Harvard University conducted the *Study of Adult Development*, which began in 1938 with

[26] Victor Frankl, *Man's Search for Meaning* (New York: Simon & Schuster, 1959) 86, 117. Available online at: https://archive.org/details/manssearchformea00fran_0/page/n5/mode/2up?q – Tr.

[27] Quoted in Frankl, *Man's Search*, 9. – Tr.

[28] Frankl, *Man's Search*, 49. – Tr.

the analysis of 700 young men. The study followed these young people throughout their lives, monitoring their mental, physical and emotional state. The study is still in progress, with over a thousand men and women who are children of the original participants. The current director of the study is psychiatrist Robert Waldinger.

In an interview with the BBC, Dr. Waldinger revealed that the study has so far yielded some important conclusions, but the key to staying happy and healthy throughout life is the quality of our relationships. The American psychiatrist states:

> What we found is that, in the case of people who are more satisfied in their relationships, more connected to others, their bodies and brains remain healthier for longer.[29]

When asked about the influence of money, Waldinger stated that, according to the lengthy study, from a financial level (when our needs are met), increasing income does not necessarily bring happiness.

Participant follow-up included questionnaires, medical records, blood tests and even DNA analyses. The study's impact led Waldinger to reflect more deeply on his own relationships and invest time in them, realizing that this brings him greater happiness. He emphasized the importance of being present and offering undivided attention to others, establishing meaningful connections. Conflicts can sap energy and health, so he encouraged building healthy bonds, even with people with whom we have disagreements.

Thus, in today's world, inundated with messages that exalt wealth and consumption, it is essential to remember that interpersonal relationships and the exercise of love are the basis for developing a full and healthy life.

Unfortunately, what we are witnessing today is a time when our values are increasingly being questioned, as stated in the Gospel of Matthew 24:12: "... because of the increase of wickedness, the love of many will grow cold..." and this increase reaches us through futile passions and human addictions, or even through the psychic anesthesia promoted by so many distractions of the modern world, such as spending many hours of our days on social media or other unproductive pastimes that cause us to lose sight of our incarnational projects (established before our rebirth).

For this reason, we always need to pause during our busy days and, for a few moments, strive to remember what those who came before us had to advise us

[29] Interview of February 28, 2023. – Tr.

about life, about living well. What did Socrates, Plato, Seneca, Marcus Aurelius, and especially **Jesus Christ** teach us?!

In this sense, the message of the Spiritist Doctrine appears as crystal clear water, which, bringing together all the philosophical, scientific and religious precepts already studied, quenches humanity's thirst for its reconnection with the Divine, combating materialism at its core.

The Doctrine clarifies the immortality of the soul, the cause of our deep sentiments and all the processes of love we need to develop to give meaning to our lives.

The Spiritist approach is fundamentally consoling and a tireless defender of life because it teaches that the pathway of spiritual progress is closely linked to the practice of loving one's neighbor and practicing charity; in addition, it demonstrates that our loved ones live forever, even those who departed before us, and that we are all involved in a complex course of growth, making every second on earth a precious learning opportunity.

There is no denying the fact that some human experiences bring us closer to the Divine Laws. The completeness that surrounds us with each smile we receive or each hope kindled in the hearts of those who suffer, enables us to feel at least a hint of complete happiness and meaning in life. The joy of providing someone else with a moment of happiness is much more rewarding than smiling by oneself.

In this context, I recall a recent account from a patient who emotionally recounted a brief encounter that completely changed his psychological state. He stated that he had been going through a difficult time in his life, and on a particular day, certain events had left him particularly distressed. It was then that, on his way to work, he observed (from inside his comfortable, air-conditioned car) a woman of approximately 60 years old searching through a pile of scrap metal for something that could be useful to her, under the scorching 1:00 p.m. sun of Belém do Pará. At that moment, feeling sorry for the woman, who looked to be about his mother's age, he stopped the car and asked if he could help her. He then gave her the amount he had in his wallet at that moment: R$100 (one hundred reals).

The woman welcomed him kindly and blessed him. After he had driven a short distance, he glanced in the rearview mirror and saw the woman kneeling in tears, arms outstretched, looking up at the heavens, thanking God in a gesture of pure humility and gratitude.

That meeting caused the patient to reflect deeply, redefining his inner concerns and making him feel like the real and greatest beneficiary of that day's experience.

Finally, how can we not mention what, for many, represents the pinnacle of human experience: the birth of a child?

For those who have had the opportunity to live such an experience and surrendered themselves to it, they certainly experienced a spectacular moment of intimate connection with the Divine, feeling the pulse of the greatest force in the Universe: genuine love with no hidden agenda.

Following the parental experience, the first smile, the first words, the first steps and the entire natural evolutionary sequence of the human species, becomes a sublime act of happiness and meaning in the lives of loving parents, enabling them to understand co-creation as one of the great purposes of their existence.

Therefore, we advocate here the non-negotiable idea of beauty and the need to preserve Human Life.

Even in the face of difficulties, even in the face of pain, let us remain with the purity of the children's response… "and Life, it is beautiful and it is beautiful."

9.3 Cases of "non-abortions": when life spoke louder

In the 1920s, specifically in mid-1928, a surprising event occurred in the life of Mrs. Elsa Bolaños Cacho. Unaware that she was pregnant, she took a medicine containing quinine. This substance, commonly used to treat infectious diseases and relieve muscle spasms, is widely discouraged during pregnancy due to the risk of congenital malformations. Faced with this delicate situation, doctors suggested abortion as a preventive measure.

However, Mrs. Cacho vehemently refused; even facing adversity and uncertainty, she remained firm in her conviction. Months later, to everyone's surprise, a healthy son was born to her, bringing joy and hope to the family.

This son grew up to become an unforgettable icon of television and humor in Mexico, captivating the hearts of people in countries around the world. His name was Roberto Gomes Bolaños, *immortalized as the beloved character El Chavo.*

In 1984, Angela Silva, a mother of two, became pregnant for the third time. The family was experiencing serious financial difficulties and saw abortion as a way to minimize the hardships of having another child in the household. At a difficult time, after considering abortion, Angela had an emotional and profound conversation with her grandfather, who convinced her to abandon the idea and carry her pregnancy to term.

Years later, her son, Brazilian soccer defender Tiago Silva, became the mainstay of the entire family and, through a successful sports career, made his mark on history, even becoming captain of his country's national soccer team.

Dolores Aveiro, in her biography entitled "Mãe Coragem" (Mother Courage), recounts that in 1984 she made two attempts to abort what would be the youngest child in the family. First, she sought out a doctor, who refused to perform the procedure on grounds that she was young, healthy and had no impediment to raising another child. Desperate and guided by a neighbor, Dolores resorted to a home remedy: she drank warm beer and ran for miles and miles in an attempt to terminate the pregnancy. However, the most she achieved was a bad upset stomach.

Dolores did not want to have another child due to the complicated life she led on the island of Madeira, a Portuguese archipelago located off the west coast of Africa. In addition to financial difficulties, she also had to deal with her husband José Diniz Aveiro's alcoholism. After several failed attempts, one of the greatest soccer players of all time was born in 1985. He has been voted the world's best player five times and is one of the wealthiest individuals in soccer history. His name: Cristiano Ronaldo.

"I wanted to meet [her] mostly to see if she was okay and to thank her, because I'm glad I didn't end up as an abortion," he said. "She was 23 and she went through a lot to have me."[30] Such is a personal statement by one of the greatest icons of the last century: entrepreneur, inventor and revolutionary magnate in the computer field, Steven Jobs, as he tells the story of how he was given away by his mother shortly after his birth.

Singer Celine Dion, her mother's fourteenth pregnancy, was almost never born. Her mother was persuaded by a priest to give up on getting an abortion. Consequently, she gave birth to one of the greatest Canadian singers of the last century.

The phenomenal tenor Andrea Bocelli also survived the strong medical recommendation for an abortion when his pregnant mother developed acute appendicitis. Even knowing the possibility of the child being born with some infirmity, she refused to abort and heroically carried the pregnancy to term. To date, Andrea Bocelli has won five BRIT Awards and three Grammys; he has recorded nine complete operas (including: *La Bohème, Il Trovatore, Werther* and *Tosca*), as well as several classical and popular albums, having sold more than 70 million copies worldwide and moved thousands of people through his art.

---

[30] Walter Isaacson: *Steve Jobs* (Simon & Schuster, 2011), 233. – Tr.

Ana Alvez Franco's thirteenth pregnancy took her by surprise, considering she was already experiencing the first signs of menopause. With her body weakened after twelve births and three miscarriages, she received medical advice to terminate the pregnancy in an attempt to avoid potentially serious health problems. The doctor advised her with the following words: "Dona Ana, let's have an abortion! You've already fulfilled your mission as a mother with the twelve children God had given you. This child will put your life at risk! You're in no health condition to survive!" Given the doctor's arguments, anyone would probably have considered the proposal. Furthermore, Ana's family was facing numerous financial difficulties and would face even more if another child were born. Even so, against what many would consider an obvious solution, Ana turned to the doctor and replied: "No, doctor! I will not abort my child! If I die in the hope of giving life to a child, it will be an honor for me!" Nine months later, Divaldo Pereira Franco was born, living proof of unconditional love and maternal sacrifice.

Divaldo Franco[31] is a true apostle of Spiritism. Of his more than ninety years, seventy were devoted to the Spiritist cause and to the marginalized children of his hometown, Salvador. He is recognized as one of the greatest Spiritist mediums and speakers of all time.

His resume also reveals a great humanist with more than 600 children adopted in circumstances of extreme need, providing love and dignity to their lives.

He has delivered over 11,000 lectures in more than 2,000 cities throughout Brazil and in 62 countries, and has given over 1,100 radio and television interviews on more than 450 stations. He has received over 700 honors from cultural, social, religious, political and governmental institutions.

As a medium, he has published over 200 books, with over 8 million copies sold, featuring 211 spirit authors.

In 1952, together with Nilson de Souza Pereira, he founded *Mansão do Caminho (Mansion of the Way)* in Salvador, Bahia, an institution that welcomed and educated children under the system of substitute homes. In 20 such homes, he raised over 600 children, who are now emancipated, most of whom have established families.

Today, *Mansão do Caminho* is an impressive educational complex covering 83,000 square meters and comprising 50 buildings, serving 3,000 children and young people from low-income families in one of the lowest-income

[31] Divaldo Pereira Franco, May 5, 1927 – May 13, 2025. – Tr.

neighborhoods of Salvador. The complex hosts a range of socio-educational activities. To date, more than 30,000 children have participated in the various courses and workshops offered by *Mansão do Caminho*. The work is basically maintained through the sale of mediumistic books, and the lectures and seminars given by Divaldo.

The extraordinary stories described above have in common beautiful examples of women who resisted pressures of various kinds and decided to give birth to their children, even in the face of pronounced difficulties and negative prognoses.

Models of determination and unconditional maternal love, they defied the odds, proving that even when circumstances seem unfavorable, life is worth living.

And life, and what is life?
Tell me, my brother
It is the beating of a heart
It is a sweet illusion
And life
Is it a wonder or is it suffering?
Is it joy or sorrow?
What is it? What is it, my brother?
There are those who say that our life
Is nothing in the World
It is just a drop in time
That does not even last a second
There are those who say it is divine
A profound mystery
A breath of the Creator
In an attitude replete with love
You say it is a struggle and a pleasure
He says life means living
She says it is better to die
Because she is not loved and the word means suffering
I just know that I trust the girl
And in the girl I put the strength of faith
We are the ones who make life

As one can, or as one wishes
Always wanted!
No matter how wrong it may be
No one wants death
Only health and fortune
And the question rolls around
And the head shakes
I shall take the purity
Of the children's response
And life, it is beautiful
It is beautiful

(Gonzaguinha)

CHAPTER X
FINAL THOUGHTS

The preceding chapters have sought to expose and defend, above all, the inestimable value of human life.

The conclusion that may be drawn from the work is that, as a society, we can join forces, hopeful that in the near future the obvious will not have to be defended, and that incipient lives will not have to be "bartered."

We have seen that, funded by a veritable billion-dollar industry, the pro-abortion movement has been actively working in recent decades to transform the unacceptable termination of a defenseless, vulnerable human life into something acceptable and even a "woman's fundamental right."

This book was born from the desire that such a scenario must and can be combated with information and counterarguments, as well as support and assistance to mothers who find themselves in vulnerable situations (physical, psychological and spiritual).

What seems unreasonable, under any circumstances, is accepting the narrative that minimizes the life of the developing fetus, transforming it into something non-human, reducing it to a cluster of cells, but, at the same time, defending the existence of turtle eggs or endangered embryos from other animal species.

It has been demonstrated that abortion movements deny the basic foundations of biology and genetics, as well as the individuality present beginning at conception and evidenced even in the presence of the fetal psyche.

It has also been shown that the situation becomes direr when we consider abortion procedures in the equation. These are shocking and cruel acts against fetuses, culminating in the insane brutality of certain places allowing abortion even when the baby is ready for life outside the womb. In such cases, we would argue that the best solution would be to deliver the baby and hand it over to competent support agencies, rather than sacrificing it.

Another statement disconnected from reality was also studied: that abortion would protect the woman. In fact, we have seen that science proves exactly the opposite, when we observed, for example, the alarming increase in maternal mortality in cases of abortion – up to ten times higher compared to women who did not terminate their pregnancies. And if that were not enough, the physical and psychological repercussions can be devastating for the woman undergoing the procedure, including increased rates of breast cancer and psychiatric syndromes (especially anxiety disorders and depression). Therefore, we conclude that there is no such thing as safe abortion!

We also hope we have sufficiently demonstrated that the voluntary termination of pregnancy brings no benefits to women or to fetuses; hence, defending the lives of the unborn means defending the lives of women.

Indeed, when we understand that we are much more than a cluster of cells and that our minds are not contained in the electrochemical manifestations of the brain, we are indeed experiencing matter, yet we are not reduced to it. The connection between science and spirituality proposes a new scientific paradigm, one that takes into account the non-material realm and everything that involves it.

In this way, the spirit-matter duality begins to be considered, and the simple fact that it is a hypothesis should already have a definitive impact on those who defend abortion, considering the possibility of the existence of a spirit-related life linked to the embryo beginning at conception.

As exemplified in chapter VI, the segments of psychiatry/psychology that utilize the tool of past life memory regression, the vast documented evidence of near-death experiences, mediumistic evidence in scientific publications, etc., make way for a new understanding of the human experience.

In light of this new paradigm, the Spiritist Doctrine gains even more relevance because it proposes a highly lucid and mature vision in relation to the dualistic concept of spirit and matter, presenting reincarnation and the possibility of communication with the spirit realm as its fundamental pillars. It aligns with scientific currents that work with regression of past life memories and with parapsychology studies that encompass mediumistic issues within serious analysis methodologies.

From a Spiritist perspective, we understand that Divine mercy, through reincarnation, always allows us to start over and repair our wrongs on the pathway of the good, as pointed out in [I Peter] where it states "...love covers a multitude of sins."

Finally, understanding that we all need to shed light on issues surrounding abortion, I hope to have contributed to important debates and reflections on this very serious topic, as humanity needs to awaken from the deep slumber that has, for millennia, kept us tied to the barbaric and cowardly acts of extermination of innocent lives, hindering the evolutionary march of thousands of spirits.

GRATITUDE

Thank you so much, Lord!
Thank you so much for what you have given me.
Thank you so much for what you give me.

Thank you for bread, for life, for the air, for peace.
Thank you so much for the beauty my eyes
behold on the altar of nature.
Eyes that gaze at the sky, the earth, and the sea
That follow the light bird that flies
joyfully across the indigo sky
Eyes that rest on the green earth, dotted
with flowers in a thousand hues.

Thank you so much, Lord!
Because I can see my love.
But before my vision
I detect blind people walking in the darkness
who stumble in the crowd
who weep in solitude.

I pray for them and I beg you for compassion
for I know that after this struggle, in the next life, they too
will see!

Thank you so much, Lord!
For my ears, which God gave me.
Ears that hear the patter of rain on the shed
The melody of the wind in the branches of the elm
The tears that flow from the eyes of the whole world!

Ears that hear the music of the people who
come down from the hill to the square, singing.
The melody of the immortals, heard once
and never forgotten!
The melodious, melancholic voice of the cowboy.
And the pain that moans and cries in the heart of the whole world!

For my joy of hearing,
for the deaf, I want to ask you,
Because I know,
That after this time of trial, in your kingdom of love, they will hear again!

Thank you for my voice
But also for your voice
For the voice that sings
That loves, that teaches, that instills literacy,
That hums a song
And that speaks Your name with heartfelt emotion!

Before my melody
I want to pray for those who suffer from aphasia.
They do not sing at night; they do not speak during the day.
I pray for them
For I know that after this trial, in their new life
They will sing!

Thank you, Lord!
For my hands
But also for the hands that plow
That sow, that shelter.

Hands of tenderness that free from bitterness
Hands that hold hands
Of charity, of solidarity

Hands of farewell
That remain wounded
That wipe away tears and pain!

[I pray] For the hands that play symphonies, write poetry,
Perform surgeries, psychographies!
For the hands that care for old age
For the suffering
For the lack of love!
For the hands that cradle the body
of another's child, without fear!
And for the feet that bear me, without complaint!

Thank you, Lord!
Because I can move.
Before my perfect body
I want to pray to you
Because I see on Earth
The disabled, the maimed, the paralyzed
who cannot move.

I pray for them
For I know that after this expiation
In the next reincarnation
They too will dance!

Lastly, thank you for my home.
It is so wonderful to have a home!
It does not matter if this home is a mansion, a hovel,
a shack, a nest, a grief-stricken place, a bungalow,
a house on the road, whatever it may be.

May there be within it the figure
Of maternal or paternal love
Of a wife or husband
Of a son or brother
The presence of a friend
The company of a dog
Someone to extend a helping hand!

But if I have no one to love me,
No roof over my head,
No bed to lie in,
I will not complain.
On the contrary, I shall say

Thank you, Lord!
Because I was born!
Thank you because I believe in you.
For your love, thank you, Lord!

Amélia Rodrigues
From the book Sol de Esperança,
Psychographed by Divaldo Franco

APPENDIX I

As applied in the Spiritist doctrine, the exercise of mediumship, when aligned with Christian morality, reveals the reality of the spirit realm, and proffers extremely comforting reflections, especially when we understand that, greater than the pain of losing a loved one, is the joy of reunion in that realm.

Thus, it is clear that there is no eternal suffering. We are constantly maturing as spirits, and separations (when one of the parties returns to the spiritual homeland) are always temporary.

Drawing on mediumistic psychography, we extracted three messages by spirits who went through delicate situations involving the topic of abortion, giving us important elements for glimpsing the grandeur of Divine mercy and how we are architects of our own destiny.

Below are texts from a mother, a son and a doctor, who from the spirit realm, recounted a little of their stories...

1

... My name is Madalena. I lived in the 18th century beside a road on which cattle drivers passed, in the city of Minas Gerais [Brazil]. I worked as a cook, but, out of necessity from living in extreme poverty, I also slept with men in exchange for money.

One day, I met a cattle driver and fell in love with him. From that relationship, a life was born within me.

I immediately tried to tell the news of my pregnancy to the man I loved, but he rejected me. He said he did not want to hear about the pregnancy and that if I ever approached him again, he would kill me and the baby.

At that very moment, my love turned into hatred, and overcome with rage, I decided to end the life of the child that was in my womb by having an abortion.

I underwent the procedure at about 20 weeks and ended up holding my daughter's lifeless body, which I threw away.

Years later, I discarnated and found myself in immeasurable darkness. There were many years of intense pain and suffering in the state of erraticity[32] *in the umbral.*[33]

I incarnated again in the 19th century, into a wealthy family in the city of Campos dos Goytacazes, and married a doctor.

My husband often traveled to Rio de Janeiro for work, and over time, I ended up getting involved with a local young man; I became his mistress and would meet with him every time my husband traveled. I got pregnant. In love, I made plans to end my marriage and run away with the man I loved. I planned everything. We worked out all the details, and I left home. However, he never showed up. On the appointed day and time, he left me all alone, suitcase in hand, at a crossroads. Once again, I felt the force of the hatred of unrequited love take hold of me. I got another abortion. I discovered that once again, I had denied the opportunity of life to a girl.

That day I returned home, intending to continue with the marriage. Three days later, I suffered a severe hemorrhage. I was supported throughout by my husband, who, despite having medical knowledge that it was a miscarriage, did not judge me and did everything he could to save me. Nonetheless, I discarnated.

Once again, I found myself in a long, dark period of suffering and darkness.

I incarnated yet again in the 20th century, in Belém do Pará, in a very humble neighborhood called Jabatiteua, into a very poor family in which I had 11 siblings.

I worked from a young age, helping out as a maid in a family home, where I later lived. In my youth, I met a shoemaker, with whom I fell in love and developed a relationship. From that relationship, I got pregnant, and once again, the father wanted nothing to do with the child and ordered me to have an abortion.

My employer found out what had happened and refused to allow it. She told me to carry the pregnancy to term and that after the child's birth, she would take care

---

[32] Erraticity [errant state] – The state of wandering or errant spirits, i.e., not incarnate, during the intervals between their corporeal existences. *The Mediums' Book*, Chap. XXXII, SPIRITIST GLOSSARY. – Tr.

[33] The *Umbral* is the region "situated between heaven and earth. It is the doleful region of shadows built and cultivated by the human mind, which is generally rebellious, lazy, confused and feeble…" (Francisco Candido Xavier, *Action and Reaction*, International Spiritist Council, 2010) – Tr.

of it. I struggled through the pregnancy, feeling very angry throughout the months. But I had the baby: a daughter, whom I named Esmeralda.

Esmeralda grew up, living with me in the home of the family where I worked; they gave us their total support. But at no point did I feel love for the child.

One day, when my employers traveled to Mosqueiro, I gave the girl to a couple who were moving to Rio.

When my employer returned and found the child missing, she threatened to fire me and report me to the police if she was not returned.

Out of fear, I managed to get Esmeralda back and handed her over to my employer. I was then fired.

Unemployed, I fell into prostitution, began drinking heavily, was constantly drunk, and moved to the city of São Miguel. There, I spent the rest of my days living a life tormented by poverty and illness. One day, already elderly, I was standing in front of my house when a young woman drove up in a very nice car. It was Esmeralda. She was grown up, beautiful, and had married an older man, a German, who lived in Rio.

My daughter said that even though I had abandoned her, she would not do the same to me or leave me there in that condition. She took me back to Rio and cared for me until I discarnated. Days before I discarnated, I asked my daughter for forgiveness, with all the sincerity in my heart, saying that I deeply regretted everything I had done. And she forgave me.

This time, upon discarnating, I ended up in the umbral once again, overcome by the guilt that still consumed me. But I did not stay there for long. I was quickly rescued and finally learned our story: Esmeralda had long been a great enemy, who had to be born as my daughter for the good of both of us, and I had so often denied her that opportunity. I also learned that her father, who had also rejected her in many lives, was also an enemy who needed redemption.

Nowadays, whenever I am allowed, I speak to my beloved daughter in her dreams, and we have become great friends, united by the true feeling of love.

Esmeralda is elderly mow and will discarnate soon. I eagerly await another opportunity for us to reincarnate together so that I can be her mother again, this time without the shadow of abortion, and so that I can fully exercise maternal love.

Communication received during a mediumistic meeting at the Jardim das Oliveiras Spiritist Center, on 10/04/1923, by the medium Paulo Oliveira.

2
MOMMY... PLEASE DON'T KILL ME AGAIN

...I struggled; I worked; I committed myself to getting authorization to be reborn; and you committed yourself to me; to me and to God...

How happy I was the day that you, in spirit, alongside Dad, agreed to welcome me into the intimacy of your home.

I longed for a new body that would enable me to redeem my past wrongs. I planned a future of light. Yes, my life would be marked by trials and redemptive experiences. Nevertheless, I prepared myself by trusting in your love. And at the moment I needed you most, you murdered me.

Why, mommy? Why?

When you sensed me in the sanctuary of your womb, you changed your behavior and began to torture me. Your thoughts of rebelliousness, unheard by anyone, echoed in my incipient ears, like heartbreaking screams that greatly distressed me. The cigarettes you smoked poisoned me.

Your nervousness, the result of your inconformity, resulted in real floggings.

When you decided on an abortion, a tremendous struggle occurred: you, wanting to expel me from your womb; I, fighting to stay.

Why did you close your ears to the voice of conscience that asked you for compassion and serenity? Why did you numb your feelings to the point of forgetting that I would bring a world of blessing and joy to you?

I would have been an obedient and loving son. I would have brought the means to sustain you in the final years of your presence on earth. However, you refused. Look at the consequences: I am tormented by not being able to be born, and you are sick, sad and restless. Your mind is tormented by affliction and your dreams are filled with nightmares.

Why, mommy, why didn't you let me be reborn?

"It's still early," you thought. "I want to enjoy life, go out, have fun, travel. Children, only later."

No child arrives at the wrong time, however.

The laws of life are wise, and no one is born by chance. However, because of the great love I have for you, I am asking God's mercy for you. I even dared to intercede for you to attain the blessing of rebalance so that, in the near future, we will be together, I in your womb and you, as always, in my heart; I nourishing myself with your vitality, and you strengthening yourself in the greatness of my purest sentiments.

Mommy, please don't repeat your premeditated act.

When you feel someone knocking on the door of your heart again, it will be me, the rejected son, who has come back to live and help you be happy.

Mommy, don't forget me; don't abandon me; don't expel me; don't kill me again; I need to be reborn..."

From the book *O que Dizem os Espíritos sobre o Aborto* (What the Spirits Say about Abortion) (Federação Espírita Brasileira, 2007), Chap. XII – *Reação dos Espíritos vitimados* (Reaction of Spirit Victims).

3

I was born in Belém, State of Pará, Brazil, into a very poor family. My mother raised us by being a laundry woman. I was one of ten children. I helped my mother by working at the street market, but I continued my education. Having always enjoyed studying, I chose to study medicine, and I managed to get into the Faculdade de Medicina do Estado do Pará (Pará College of Medicine). I was a good student.

Starting in my third year, I accepted a job offer because I needed money when my wife got pregnant (I had married an old girlfriend and childhood neighbor, a nurse technician). We lived together in an annex to my mother's house, with little money and a lot of hardships.

The work invitation had come from a certified physician who needed support with his procedures at an illegal abortion clinic. Fixated on our hardships and only thinking about the money, I began helping him.

After almost two years of working at the clinic and demonstrating great surgical skill, I progressively took on more work, increasing the clinic's overall productivity.

That was when I realized I could make a lot of money on my own. I dropped out of medical school (in my fifth year) and dedicated myself entirely to performing abortions. I bought a property near the city museum (Emílio Goeldi) and turned it into my clandestine clinic, with the support of my wife, who helped me with the procedures. We amassed a fortune and bought an apartment and a beach house. I worked from Sunday to Sunday, focused solely on money; my mind felt nothing – until I lost a four-year-old son who was run over. When I stopped to feel the pain of having lost my son and as his absence grew stronger, I began to think about how many people did not want their children, while I was suffering with the absence of my own. At that moment, I became deeply conflicted and no longer wanted to go to the clinic. I was under a lot of pressure from my wife and other older children because no more money was coming in. I started drinking and despairing at the age of fifty-two, having worked for decades, so I fell into a deep depression. And one day, walking the streets of Belém, drunk, I was killed after being hit by a bus.

Upon arriving in the spirit realm, I had no peace for almost twenty years, persecuted and condemned by so many victims of whom I had been the executioner. And what hurt me most was that I could not find my son.

My regret was immense. I saw that I had had all I needed to have had a career as an honest doctor, because I had skills as a surgeon; I could have done much good.

There was no peace in my conscience, but over time I began to pray, work and understand that I needed to reincarnate. I will return without either hand, blind and autistic, in a very poor family. That is what I chose to try to temporarily forget this weight on my conscience and I ask Jesus and all those I have wronged to forgive me. And I ask you to pray for me... that is my story.

Communication received during a mediumistic meeting at the Jardim da Oliveiras Spiritist Center, on 05/31/1923, by the medium Paulo Oliveira.

APPENDIX II
"LET US UNITE IN THE DEFENSE OF LIFE"

My Children!
May Jesus bless us!

Considered from any perspective, life is a gift from God that no one may disturb. On the evolutionary scale, all sentient beings have a plan developed that seeks fulfillment, perfection, which is their ultimate goal.

Preserving life, in all its expressions, is an inalienable duty that human consciousness assumes in the development of its evolution.

When someone raises a club to deliberately interrupt the cycle of life, they become a new Cain, condemning themselves to a guilty conscience, experiencing remorse, leading them, sooner or later, to the need to purify themselves, rehabilitating themselves by swimming in rivers of tears.

Therefore, Christian Spiritists, understanding the lofty meaning of life, are rising up to defend it wherever it appears, especially human life—an advanced stage in the process of enlightenment of the Being, in the search for its complete cosmic consciousness.

Inspired by the higher spirit realm, Christ's workers are rising up today to proclaim not only the right to life of those who are in the womb and have the right to be born, but also of those who are about to discarnate, yet have the right to remain, until their last breath, in a degenerating organism; of unrepentant criminals, who can finally repent and become useful instruments to the community they have harmed, or the mentally, spiritually and morally tormented who, without the strength to face the struggle, opt for the wrongful solution of suicide, plunging into the unfathomable abyss of darkness and suffering.

Not only defending this right to life, but also respecting all lives, however they may be presented and wherever they may be, is the primary task of Spiritism, which can be considered a powerhouse of awesome force. If, by chance, it does not carry out the transformative operation of its members, influencing the behavior of society, it becomes a force left by the wayside, which has lost the purpose of its intended use.

Therefore, Spiritism's primary objective is the moral transformation of humankind, but if that does not occur, its message may be compared to a blessed lamp whose light, unfortunately, has gone out.

From that individual and non-transferable moral transformation, other objectives will arise, objectives which will meet collective necessities, changing the landscapes of the earth and inviting individuals to participate in the actual construction of the better world that we will soon encounter.

And where else would the energy needed for this task come from, if not from the home, that miniature society where sentiments are merged, rough edges are smoothed, and, often, as with a chisel, flaws and imperfections are removed, filing away the rough edges so that the inner light can reach the surface and expand?!

The family is the foundation of society, and it cannot be relegated to a secondary role. Living in a family with dignity and honor is to value life, taking advantage of the opportunity that God has granted the spirit so that it may grow and reach the heights for which it is destined.

It is true that many obstacles arise, creating difficulties for both undertakings.

Who, by chance, will cross a river twice in the same waters?

Facing such obstacles is the decision of the renewed Christian, who has found in Jesus the powerful strength that He used when He wanted to establish His Kingdom of love and justice in His time, which was much the same as our time, all things considered.

If individuals will arm themselves with moral courage to combat error, through their attitude of Spiritist-Christian consistency and their eminently evangelical conduct, slowly but surely the lost ground will be regained and the Kingdom of Labor, Fraternity and Love will be established on Earth.

My children, there is much darkness because the good is presented timidly, yielding to evil, which is expanding its domains through the boldness with which it is clothed.

For this very reason, may our definition be Spiritist.

If it is necessary to lose Caesar's sad coins in order to preserve the integrity of the message's content, let us trust in God, the Supreme Giver, who has never left us orphaned and will never abandon us.

Spiritism frees us from ignorance and, via the enlightening lessons of charity, encourages us to engage in social action by providing assistance and relief services. To deny the origin of inspiration, to be in collusion with the arbitrary and unjust methods of earthly politics, is the same as yielding to the passions of Caesar as the arbiter of destinies, albeit without control over lives; it would mean renouncing the name of Jesus, who is the banner of our social endeavors, and being arm in arm with the temporal power, receiving its help and supporting its arbitrariness.

Jesus said that in the world we would only have afflictions.

Hence, it would not be right to expect any other response than that of difficulty.

Thanks to the Sovereign Law, which is the Natural Law, the Law of Love, let us struggle along with the competent authorities to speak about our apostolate and demand respect for the socially renewing actions that we have been developing in the name of charity.

Let us never fear! Let us be united in defending life in a worthy spiritual family, enduring setbacks and misunderstandings. Being a Spiritist today is the same as having been a Christian yesterday.

How many times will we see our best words adulterated and turned against us?

At other times, we will face the challenges of slander, corruption, and undue accusations in new situations. We will confront intimate problems in the sanctuary of our homes, tearing at our hearts, and later on, we will suffer the insidious interference of those who take pleasure in preserving this state of affairs, tormented by their inferior wrongfulness, wounding the most intimate fibers of our sentiments.

It was for no other reason that the Master recommended love to us – love always – and prayer, my children!

Prayer is the elixir of long life that provides us with the resources to preserve uplifting values, persevering in enlightening work. And indiscriminate love for everyone, even enemies—which does not mean agreeing with their misdeeds—is imperative for us to achieve peace.

As it is true that His disciples will be known for loving greatly, it is no less true that this love—which begins within us—must expand to include them, all of them, those who cause us embarrassment and difficulties, who threaten us and bring us to tears, on both planes of life.

Therefore, trust in Jesus.

When we accept the ministry of Restored Christianity, we sign a commitment to serve selflessly until the end.

We have been successful; we have won the first battles; we have overcome the greatest difficulties before the decision. We now need to value our lives—you, in the material world, and the rest of us, in the experience of freedom from the body—to reach the final goal, singing a hymn to the victor who, apparently defeated, was planted on the cross, and whose apparent defeat was symbolized in the victory of finding himself as a "Hyphen of Light" between humankind and God, on Calvary, where all the forces of evil united for the sacrifice of the Lamb.

My children, these days have come. Be of good cheer, maintain your courage, be faithful, valuing life and living in a family with dignity, so that you may establish on earth the ideal family, whose members, bound to the Kingdom of God, are truly brothers and sisters.

May the Lord bless us!

These are the words of the workers from the Beyond who, through us, bring them to your minds and hearts.

With the usual paternal affection, the most humble servant.

Bezerra

(Psychophonic message received by the medium Divaldo Pereira Franco at the closing of the National Federative Council meeting, on 7-11-1993 in Brasília).

BIOGRAPHICAL REFERENCES

Araújo, Saulo de Freitas, 2012. "Materialism's eternal return: recurrent patterns of materialistic explanations of mental phenomena." In *Archives of Clinical Psychiatry*, (São Paulo), vol. 40, no. 3, 114-119.

Aurelius, Marcus, *Meditations*. London: Dent, 1906.

Beauregard, Mario, and Denyse O'Leary. *The Spiritual Brain*. Harper-Collins, 2007.

Beauregard, Mario, Gary E Schwartz, Lisa Miller, Larry Dossey, Alexander Moreira-Almeida, Marilyn Schlitz, et al. "Manifesto for a Post-Materialist Science." In *Explore*. Oct. 2014; 10:272-4.

Beischel, Julie, and Gary E Schwartz. "Anomalous information reception by research mediums demonstrated using a novel triple-blind protocol". National Library of Medicine - National Center for Biotechnology Information, 2006. Available at https://pubmed.ncbi.nlm.nih.gov/17234565/.

Bianchi-Demicheli, F., 2007. "Psychiatry and psychological consequences of abortion." In *Revue Médicale Suisse,* vol. 14, no. 3, 401- 402, 404; 406-407.

Brasil. Ministério da Saúde. DATASUS. [2019]. Available at: http://tabnet.datasus.gov.br/.

Brind, J. *Abortion and Breast Cancer: Recent Evidence Confirms a Robust Link*. 2015.

Broen, Anne Nordal, Torbjorn Moum, Anne Sejersted Bödtker, and Oivind Ekeberg, 2005. "The course of mental health after miscarriage and induced abortion: a longitudinal, five-year follow up study." In *BMC Medicine*, vol. 3, no. 18, 1-14. Available at: https://bmcmedicine.biomedcentral.com/articles/10.1186/1741-7015-3-18.

Broen, Anne Nordal, Torbjörn Moum, Anne Sejersted Bödtker, Oivind Ekeberg. "Reasons for induced abortion and their relation to women's emotional distress: a prospective, two-year follow-up study." National Library of Medicine - National Center for Biotechnology Information, 2005. Available at: https://pubmed.ncbi.nlm.nih.gov/15694217/.

Buluk, U., P. O. O'neill, 1980. "Health professional's perceptions of the psychological consequences of abortion." In *American Journal of Community Psychology*, vol. 8, no. 1, 67-75.

Carroll, Patrick S., 2012. "Are full term pregnancies and induced abortion separate risk factors in female auto-immune diseases?" In *Studies on Ethno-Medicine*, vol. 6, no. 2, 121-130.

Carter, Christopher David. *Science and the Afterlife Experience: Evidence for the Immortality of Consciousness.* Rochester: Inner Traditions, 2012.

Clarke, Chris. *Consciousness and No-Collapse Physics.* 2012. Available at: file:///E:/arquivos/downloads/Consciousness_and_no-collapse_physics.pdf.

Coleman, Priscilla K., 2011. Abortion and mental health: quantitative synthesis and analysis of research published 1995-2009. In *The British Journal of Psychiatry*, vol. 199, no. 3, 180-186.

Coleman Priscilla, 2011. "Abortion and Mental Health." In *The British Journal of Psychiatry.*

Committee on Gynecologic Practice. 2009. ACOG Committee opinion no. 434: "Induced abortion and breast cancer risk." In *Obstetrics and Gynecology*, vol. 113, no. 6, 1417-1418.

"Crianças com memórias de vidas passadas surpreendem pesquisadores." In *Epoch Times Brasil*, 2014. Available at https://www. epochtimes.com.br/ criancas-memorias-vidas-passadas-surpreendem-pesquisadores_73881.html.

Daling, J. R., K. E. Malone, L. F. Voigt, E. White, and N.S. Weiss, 1994. *"Risk of breast cancer among young women: relationship to induced abortion."* In *Journal of the National Cancer Institute.*

Daling, J. R., K. E. Malone, L. F. Voigt, E. White, N. S. Weiss. "Risk of breast cancer among young women: relationship to induced abortion." National Library of Medicine - National Center for Biotechnology Information, 1994. Available at: https://pubmed. ncbi.nlm.nih.gov/7932822/.

Deng, Y. H. Xu, and X. Zeng, 2018. "Induced abortion and breast cancer: An updated meta-analysis." In *Medicine,* vol. 97, no. 3, e*9613.*

Derosa, Marlon (Org.). *Precisamos falar sobre aborto: mitos e verdades.* Florianopolis: Editora Estudos Nacionais, 2018.

DiPietro, Janet A., Matthew F. S. X. Novak, Kathleen A. Costigan, Lara D. Atella, Sarah P. Reusing "Maternal Psychological Distress During Pregnancy in Relation to Child Development at Age Two." Society for Research in Child Development, 2006. Available at: https://srcd.onlinelibrary.wiley.com/doi/ abs/10.1111/j.1467-8624.2006.00891.x.

Dos Santos, Emanuel Burck. *Consciência Imortal,* 2019.

"Experiência no além: o neurocirurgião que entrou em coma e passou a acreditar em vida após a morte". Aventuras na História, 2021. Available at: https:// aventurasnahistoria.uol.com.br/ noticias/reportagem/experiencia-no-alem-o-neurocirurgiao-que-entrou-em-coma-e-passou-acreditar-em-cida-apos-morte. phtml. Almeida, Alberto, and Gustavo Machado. *Respeito à Criança,* 2015.

Fenwick, Peter. "Can Near-Death Experience Contribute to the Debate on Consciousness?" In *Exploring Frontiers of Mind-Brain Relationship.* Alexander Moreira; 2012.

Ferguson, D. M., L. J. Horwood, and J. M. Boden, 2013. "Does abortion reduce the mental health risks of unwanted or unintended pregnancy? A re-appraisal of the evidence." In *Australian & New Zealand Journal of Psychiatry,* vol. 47, no. 9, 819-827.

Fergusson, David M., L. John Horwood, Joseph M. Boden. "Abortion and mental health disorders: evidence from a 30-year longitudinal study." National Library of Medicine - National Center for Biotechnology Information, 2008. Available at: https://pubmed.ncbi.nlm.nih.gov/19043144/.

Frankl, Viktor E. *Man's Search for Meaning.* Boston: Beacon Press, 1992.

Gissler, M., E. Hemminki, and L Lönnqvist, 1996. *Suicides after pregnancy in Finland, 1987-94*: register linkage study. *BMJ*, vol.313.

Goswami, Amit. *God Is Not Dead: what quantum physics tells us about our origins and how we would live.* Charlottesville, VA: Hampton Roads Pub. 2008.

Goswami, Amit. *The Self-Aware Universe - How Consciousness Creates the Material World.* New York: Jeremy P. Tarcher/Putnam, 1995.

Haraldsson, Erlendur. "Cases of the Reincarnation Type and Mind-Brain Relationship." In book: *Exploring Frontiers of Mind-Brain Relationship.* Alexander Moreira-Almeida and Franklin Santana Santos. Springer, 2012. 215-231.

Huang, Y., X. Zang, W. Li, F. Song, H. Dai, J. Wang, X. Liu, C. Chen, Y. Yan, Y. Wang, and K. Chen, 2014. "A meta-analysis of the association between induced abortion and breast cancer risk among Chinese females." Cancer Causes Control, vol. 25, no. 2, 227-236.

Júnior, Décio Landoli. *Da Alma ao Corpo Físico.* AMEBrasil, 2014.

Kant, Immanuel, *Critique of Pure Reason.* London: Macmillan, 1952.

Kardec, Allan. *The Gospel according to Spiritism.* Federação Espírita Brasileira, 2008, 2020.

Kardec, Allan. *The Mediums' Book.* Brasilia: Federação Espírita Brasileira, 2009, 2023.

Kardec, Allan. *The Spirits' Book*. Brasilia: Federação Espírita Brasileira, 2006, 2010.

Keenan, Katherine, Emily Grundy, Michael G. Kenward, and David A. Leon, 2014. "Women's risk of repeat abortions is strongly associated with alcohol consumption: a longitudinal analysis of a Russian national panel study, 1994-2009."

Keenan, Katherine, Emily Grundy, Michael G. Kenward, David A. Leon. "Women's Risk of Repeat Abortions Is Strongly Associated with Alcohol Consumption: A Longitudinal Analysis of a Russian National Panel Study, 1994–2009." Plos One, 2014. Available at: https://journals.plos.org/plosone/article? id=10.1371/journal.pone.0090356.

Kelly, Edward F., Emily W. Kelly, Adam Crabtree, Alan Gauld, Michael Grosso, and Bruce Greyson. *Irreducible Mind*. Rowman and Littlefield Publishers Inc., 2007.

Kisilevsky, B. S., S. M. J. Hains, C. A. Brown, C. T. Lee, B. Cowperthwaite, S. S. Stutzman, M. L. Swansburg, K. Lee, X. Xie, H. Huang, H-H Ye, K. Zhang, Z. Wang. "Fetal sensitivity to properties of maternal speech and language." National Library of Medicine - National Center for Biotechnology Information, 2009. Available at: https://pubmed.ncbi.nlm.nih.gov/19058856/.

Latt, S. M., A. Milner, A. Kavanagh, 2019. "Abortion laws reform may reduce maternal mortality: an ecological study in 162 countries." In *BMC Women's Health*, vol. 19, no. 1.

Lucchetti Giancarlo , Jorge C. Daher Jr., Decio Iandoli Jr., Juliane P. B. Gonçalves, Alessandra L. G. Lucchetti, *Historical and cultural aspects of the pineal gland: comparison between the theories provided by Spiritism in the 1940s and the current scientific evidence.* Neuro Endocrinol Lett. 2013; 34(8):745-55. PMID: 24522019.

MacMahon, B., P. Cole, T. M. Lin, C. R. Lowe, A. P. Mirra, B. Ravnihar, E. J. Salber, V. G. Valaoras, S. Yuasa. "Age at first birth and breast cancer risk". National Library of Medicine - National Center for Biotechnology Information, 1970. Available at: https://pubmed. ncbi.nlm.nih.gov/5312521/.

Mainieri, Alessandra Ghinato, Julio Fernando Prieto Peres, Alexander Moreira-Almeida, Klaus Mathiak, Ute Habel, Nils Kohn. "Neural correlates of psychotic-like experiences during spiritual-trance state." National Library of Medicine - National Center for Biotechnology Information, 2017. Available at: https://pubmed.ncbi.nlm.nih.gov/28644996/.

Mari, Jair Jesus, and Christian Kieling. *Psiquiatria na Prática Clínica,* 2013.

Moon, Christine, Hugo Lagercrantz, Patricia K Kuhl. "Language learning begins in utero, study finds; Newborn memories of oohs and ahs heard in the womb." ScienceDaily, 2013. Available at: www.sciencedaily.com/releases/2013/01/130102083615. htm.

Moreira, Alexander. "Investigating the fit and accuracy of alleged mediumistic writing: a case study of Chico Xavier's letters." In *Explore*, vol. 10, no. 5, 2014. Available at https://ufjf-br.academia.edu/AlexanderMoreiraAlmeida.

Moreira-Almeida, Alexander, "Research on Mediumship and the Mind-Brain Relationship." In A. Moreira-Almeida & F. S. Santos (Eds.), *Exploring frontiers of the mind–brain relationship* (191–213). Springer Science + Business Media. https://doi.org/10.1007/978-1-4614-0647-1_10.

Moreira-Almeida, Alexander, Harold Koenig, Giancarlo Lucchetti. "Clinical Implications of Spirituality to Mental Health: Review of Evidence and Pratical Gidelines." Braz J Psychiatry. 2014 Apr-Jun; 36(2):176-82.

Moreira-Almeida, Alexander, Marianna Costa, Humberto S. Coelho. *Ciência da Vida Após a Morte*, 2023.

Mota, Natalie P., Margaret Burnett, Jitender Sareen. "Associations between abortion, mental disorders, and suicidal behaviour in a nationally representative sample." National Library of Medicine - National Center for Biotechnology Information, 2010. Available at: https://pubmed.ncbi.nlm.nih. gov/20416147/.

Nathanson, Bernard: *The Hand of God: A Journey from Death to Life, by the Abortion Doctor Who Changed His Mind,"* Washington D.C.: Regnery Pub., 2019.

National Security Study Memorandum 200, Wikipedia, 2023. Available at: https://pt.wikipedia.org/wiki/National_Security_Study_Memorandum_200.

Novaes, Paulo Batistuta. *Temas de Valorização da Vida,* 2019.

Parnia, S. *What Happens When We Die? A Groundbreaking Study into the Nature of Life and Death.* Hay House, 2007.

Peres, Julio F. P., and Andrew Newberg. "Neuroimagem e mediunidade: uma promissora linha de pesquisa." In *Espiritualidade e Sociedade*, 2013. Available at https://www.espiritualidades.com. br/Artigos/P_autores/PERES_Julio_et_ NEWBERG_Andrew tit_Neuroimagem_e_mediunidade_uma_promissora_ linha_de_pesquisa.htm..

Peres, Julio Fernando, Alexander Moreira-Almeida, Leonardo Caixeta, Frederico Leao, Andrew Newberg. "Neuroimaging during trance state: a contribution to the study of dissociation." National Library of Medicine - National Center for Biotechnology Information, 2012. Available at: https:// pubmed.ncbi.nlm. nih.gov/23166648/.

Piontelli, Alessandra. *From Fetus to Child (The New Library of Psychoanalysis).* Routledge, 1st edition, April, 1992.

Sanger, Margaret: o controverso legado da 'mãe' do controle de natalidade, BBC News Brasil, 2023. Available at: https://www. bbc.com/portuguese/ geral-64235036.

Stevenson I. *Twenty Cases Suggestive of Reincarnation*: Second Edition, Revised and Enlarged. Charlottesville: University of Virginia Press, 1980.

Stevenson, Ian. *Where Reincarnation and Biology Intersect*, Westport: Praeger, 1997.

Thorp, J. M., K. E. Hartmann, and E. Shadigian, 2003. "Long-term physical and psychological health consequences of induced abortion: review of the evidence." In *Obstetrical & Gynecological Survey*, vol. 58, no. 1, 67-79,.

Van Ditzhuijzen, J., M. Ten Have, R. De Graae, C. H. C. J. Van Ninatten, and W. A. M. Vollebergh, 2018. "Long-term incidence and recurrence of common mental disorders after abortion. A Dutch prospective cohort study." In *Journal of Psychiatric Research*, vol. 102, 132-135.

Van Lommel, Pim. *Consciousness beyond Life*. HarperCollins, 2010.

Van Lommel, Pim, Ruud van Wees, Vincent Meyers, Ingrid Elfferich, *Near-death experience in survivors of cardiac arrest: a prospective study in the Netherlands*. PubMed – NCBI

Volkov, V. G., N. N. Granatovich, E. V. Survillo, L. V. Pichugina, and Z. S. Achilgova, 2018. "Abortion in the Structure of Causes of Maternal Mortality." In *Revista Brasileira de Ginecologia e Obstetrícia*, vol. 40, no. 6, 309-312.

Vukelic, J., A. Kapamadzila, and B. Kondic, 2010. "Investigation of risk factors for acute stress reaction following induced abortion." In *Medicinski Pregled*, vol. 63, nos. 5-6, 399-403.

Wambach, Helen. *Life Before Life: Is There Life Before Birth? 750 Cases of Hypnosis*. White Crow Productions Ltda, 2020.

Williams Kelly, Emily, and Dianne Arcangel. "An investigation of mediums who claim to give information about deceased persons." National Library of Medicine - National Center for Biotechnology Information, 2011. Available at https://pubmed.ncbi.nlm. nih.gov/21206241/.

Xavier, Chico and Hércio M. C. Arantes. *Coleção Cartas Psicografadas*. Editora Instituto de Difusão Espírita, 2011.

Xavier, Francisco Cândido, and Waldo Vieira. *Evolution in Two Worlds*. Federação Espírita Brasileira, 2013. (Series: *Life in the Spirit World*, dictated by André Luiz).

Xavier, Francisco Cândido. *On the Way to the Light*. Federação Espírita Brasileira, 2011.

Xavier, Francisco Cândido. *Pensamento e Vida*. Federação Espírita Brasileira, 2009.

Xavier, Francisco Cândido. *The Way, the Truth and the Life*. Federação Espírita Brasileira, 2012.

Xavier, Francisco Cândido. *Vida e Sexo*. Federação Espírita Brasileira, 2010.

Zafran, N.; Musa, M.; Zuarez-Easton, S.; Garmi, G.; Romano, S., 2017. *Risk of preterm birth and low birthweight following consecutive surgical and medical abortions.* In *Archives of Gynecology and Obstetrics*, vol. 296, no. 4, 763-769.

FÁBIO C. ARAÚJO JR.

Specialist in Radiology, full member of the Brazilian College of Radiology.

Postgraduate degrees in Psychiatry and training in psychotherapy and memory regression.

Director of the Medical-Spiritist Association of Pará; volunteer at the Jardim das Oliveiras Spiritist institution, where he works as a speaker and coordinator of a mediumistic group.

www.ingramcontent.com/pod-product-compliance
Lightning Source LLC
LaVergne TN
LVHW090423210926
848480LV00002B/662

* 9 7 8 1 9 5 3 6 7 2 8 5 8 *